Relativistic Effects in Interatomic Ionization Processes and Formation of Antimatter Ions in Interatomic Attachment Reactions

Andreas Jacob

Relativistic Effects in Interatomic Ionization Processes and Formation of Antimatter Ions in Interatomic Attachment Reactions

Springer Spektrum

Andreas Jacob
Düsseldorf, Germany

This dissertation was accepted by the Faculty of Mathematics and Natural Sciences at Heinrich Heine University Düsseldorf in July 2023.

ISBN 978-3-658-43890-6 ISBN 978-3-658-43891-3 (eBook)
https://doi.org/10.1007/978-3-658-43891-3

This Springer Spektrum imprint is published by the registered company Springer Fachmedien Wiesbaden GmbH, part of Springer Nature.
The registered company address is: Abraham-Lincoln-Str. 46, 65189 Wiesbaden, Germany

In loving memory of my grandfather Siddiq Abbasi

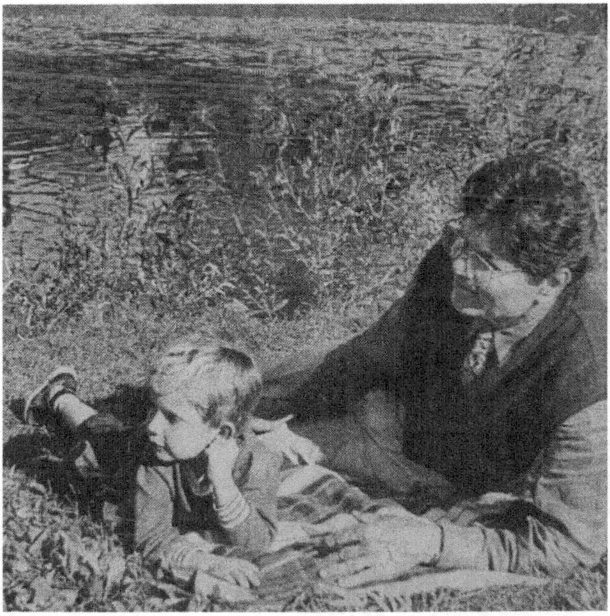

Acknowledgements

First of all, I would like to thank Dr. habil. Alexander B. Voitkiv and Prof. Dr. Dr. Carsten Müller for their great guidance and support during the years of my PhD studies as well as during my Bachelor and Master studies.

I want to thank Prof. Dr. Reinhold Egger for reviewing this thesis. In addition, I would like to thank PD Dr. Götz Lehmann for his mentorship throughout my PhD studies. Clearly, the last years would not have been so nice and pleasant without my colleagues at the institute. Therefore, many thanks to Alexandra Eckey, Dr. Lars Reichwein, Dr. Alina Golub, Dr. Fiona Grüll, Marko Filipovic and Dr. habil. Selym Villalba-Chávez. Furthermore, I want to thank Mrs. Elvira Gröters, Mrs. Ute Birkenkamp-Pickshaus and Mr. Eugen Braun for their support at the institute.

And finally, I deeply appreciate the company of my family on my journey. From the bottom of my heart, I thank my girlfriend Larissa Mieth and her mother Slawa Mieth, my parents Anja and Christian Jacob, my brother Jan-Niklas Jacob, and my grandparents Brigitte and Siddiq Abbasi.

Abstract

Interatomic energy transfer phenomena caused by efficient long-range electron correlations are among the most studied processes in atomic physics today. They can provide important insights into the interaction dynamics of single atomic species with their environment in more complex systems. For this reason, such processes are also of great interest to various other areas of physics like plasma physics, astrophysics, biophysics, and physical chemistry.

Part I of this thesis covers fundamental aspects of two interatomic ionization processes relying on the efficient transfer of electronic energy via long-range electron correlations between two spatially well-separated atomic species with an emphasis on the influence of relativistic effects on ionization. The first process is two-center impact ionization, proceeding in a weakly bound diatomic system bombarded by charged particles and involving impact excitation of one atom with its subsequent decay via efficient energy transfer to the other atom that ionizes it. This process was already considered for nonrelativistic electron impact, where it can substantially enhance total electron emission. In this thesis, two-center impact ionization is further developed by considering the impact of relativistic bare ions and by including the relativistic retardation effect, accounting for the finite propagation of the electromagnetic interaction between the atoms. We show that two-center ion impact ionization can significantly enhance total electron emission and relativistic effects caused by a high collision velocity can greatly influence the angular distribution of emitted electrons while retardation effects are mostly negligible. The second process is two-center resonant photoionization (2CPI) occurring in a diatomic system exposed to a weak laser field and involving resonant photoexcitation of one atom with its consequent decay via efficient energy transfer to the other atom which ionizes it. This process, whose high efficiency in weakly bound systems was confirmed in experiments with Ne–He dimers and Ar-Ne clusters, was studied theoretically in slow atomic collisions when the interatomic energy transfer is driven by the exchange of virtual photons. In this thesis, we extend the theory of collisional 2CPI by including the relativistic retardation effect, enabling the energy transfer to occur also via the exchange of real photons that dramatically increases the effective interaction range. Our results show that such an approach to collisional 2CPI can profoundly modify this process and strongly enhance its reaction rate.

Part II of this thesis deals with an application of interatomic energy transfer in antimatter physics, investigating attachment mechanisms for the production of substantial amounts of the positive ion of antihydrogen $\overline{\mathrm{H}}^{+}$ in view of experiments on the free-fall of antihydrogen $\overline{\mathrm{H}}$ currently planned at CERN. We perform a comparative study of various radiative and nonradiative attachment mechanisms for the formation of $\overline{\mathrm{H}}^{+}$, where special focus lies on those mechanisms driven by the efficient transfer of positronic energy via long-range positron-electron correlations in systems of antimatter embedded in matter. In the process of two-center dileptonic attachment (2CDA), a positron incident on $\overline{\mathrm{H}}$ is attached to $\overline{\mathrm{H}}$ by resonant energy transfer to a neighboring (matter) atom, which gets excited and subsequently relaxes through spontaneous radiative decay. In the process of electron-assisted three-body attachment (3BAe), a free positron and electron are incident on $\overline{\mathrm{H}}$ and the positron is attached to $\overline{\mathrm{H}}$ via efficient energy transfer to the electron with consequent increase of its kinetic energy. Our results imply, in particular, that for relatively low positron energies $\lesssim$0.1 eV (typical of current antimatter experiments) 3BAe strongly outperforms 2CDA whereas at larger energies $\simeq$ 1 eV 2CDA can greatly dominate 3BAe.

Zusammenfassung

Interatomare Energietransferphänomene, verursacht durch effiziente langreichweitige Elektronenkorrelationen, gehören heute zu den meistuntersuchten Prozessen der Atomphysik. Sie können wichtige Einblicke in die Wechselwirkungsdynamik einzelner atomarer Spezies mit ihrer Umgebung in komplexeren Systemen liefern. Daher sind solche Prozesse auch für verschiedene andere Bereiche der Physik wie Plasmaphysik, Astrophysik, Biophysik und physikalische Chemie von großem Interesse.

Teil I dieser Arbeit befasst sich mit grundlegenden Aspekten zweier interatomarer Ionisationsprozesse, die auf der effizienten Übertragung elektronischer Energie über langreichweitige Elektronenkorrelationen zwischen zwei räumlich getrennten atomaren Spezies beruhen, wobei ein Schwerpunkt auf dem Einuss relativistischer Eekte auf die Ionisation liegt. Der erste Prozess ist die Zwei-Zentren-Stoßionisation, die in einem schwach gebundenen zweiatomigen System stattndet, das von geladenen Teilchen beschossen wird und welche die Stoßanregung eines der Atome mit seiner anschließenden Abregung durch effiziente Energieübertragung auf das andere Atom, das dadurch ionisiert wird, beinhaltet. Dieser Prozess wurde bereits für Stöße mit nichtrelativistischen Elektronen betrachtet, wo er die Gesamtelektronenemission erheblich steigern kann. In dieser Arbeit wird die Zwei-Zentren-Stoßionisation weiterentwickelt, indem Stöße mit relativistischen Ionen sowie der Retardierungseffekt, welcher die endliche Ausbreitung der elektromagnetischenWechselwirkung zwischen den Atomen widerspiegelt, berücksichtigt werden. Wir zeigen, dass die Zwei-Zentren-Ionen-Stoßionisation die Gesamtelektronenemission deutlich steigern kann und dass relativistische Eekte, verursacht durch eine hohe Stoßgeschwindigkeit, die Winkelverteilung der emittierten Elektronen stark beeinussen können, während Retardierungseekte weitgehend vernachlässigbar sind. Der zweite Prozess ist die resonante Zwei-Zentren-Photoionisation (2CPI), die in einem diatomaren System auftritt, das einem schwachen Laserfeld ausgesetzt ist, und welche die resonante Photoanregung eines der Atome mit seiner anschließenden Abregung durch effiziente Energie übertragung auf das andere Atom, welches daraufhin ionisiert wird, einschließt. Dieser Prozess, dessen hohe Effizienz in schwach gebundenen Systemen in Experimenten mit Ne-He Dimeren und Ar-Ne Clustern bestätigt wurde, wurde theoretisch in langsamen atomaren Stößen untersucht, wenn der interatomare Energietransfer durch den Austausch virtueller Photonen erfolgt. In dieser Arbeit erweitern wir die Theorie von 2CPI in Stößen, indem wir den relativistischen Retardierungseekt berücksichtigen, sodass der Energietransfer auch durch den Austausch realer Photonen auftreten kann, was die eektive Wechselwirkungsreichweite dramatisch erhöht. Unsere Ergebnisse zeigen, dass ein solcher Ansatz für 2CPI in Stößen diesen Prozess erheblich modizieren und seine Reaktionsrate stark erhöhen kann.

Teil II dieser Arbeit umfasst eine Anwendung von interatomarem Energietransfer in der Antimateriephysik und untersucht Bindungsmechanismen für die Erzeugung wesentlicher Mengen des positiven Antiwasserstoffions $\overline{\mathrm{H}}^+$ im Hinblick auf die derzeit am CERN geplanten Experimente zum freien Fall von Antiwasserstoff $\overline{\mathrm{H}}$. Wir führen eine vergleichende Studie verschiedener radiativer und nichtradiativer Bindungsmechanismen für die Bildung von $\overline{\mathrm{H}}^+$ durch, wobei der Fokus auf jenen Mechanismen liegt, die durch den effizienten Transfer positronischer Energie über langreichweitige Positron-Elektron-Korrelationen in Systemen von in Materie eingebetteter Antimaterie ermöglicht werden. Bei dem Zwei-Zentren dileptonischen Einfang (2CDA) wird ein Positron, das auf $\overline{\mathrm{H}}$ auftrit, an $\overline{\mathrm{H}}$ gebunden, indem Energie resonant auf ein benachbartes (Materie-)Atom übertragen wird, welches daraufhin angeregt wird und durch spontanen radiativen Zerfall relaxiert. Bei dem elektronenunterst ützten Dreikörpereinfang (3BAe) treen ein freies Positron und Elektron auf $\overline{\mathrm{H}}$ und das Positron wird durch effiziente Energieübertragung auf das Elektron an $\overline{\mathrm{H}}$ gebunden, was mit einer Erhöhung der kinetischen Energie des Elektrons einhergeht. Unsere Ergebnisse implizieren insbesondere, dass für relativ niedrige Positronenenergien $\lesssim 0.1$ eV (typisch für aktuelle Antimaterieexperimente) der 3BAe Prozess den 2CDA Prozess deutlich übertrit, wohingegen bei größeren Energien $\simeq$ 1 eV der 2CDA Prozess den 3BAe Prozess stark dominieren kann.

List of Publications

The publications that resulted from this thesis work are listed below, including a brief account of the contributions of the thesis' author A. J. to each publication:

1. **Formation of $\overline{\mathrm{H}}^{+}$ via radiative attachment of e^{+} to $\overline{\mathrm{H}}$**

In this project, radiative attachment of a positron to antihydrogen in order to form the positive ion of antihydrogen $\overline{\mathrm{H}}^{+}$ via (i) spontaneous radiative attachment, (ii) (laser-)induced radiative attachment and (iii) two-center dileptonic attachment is studied. The contribution of A. J. includes the analytical calculation of the $\overline{\mathrm{H}}^{+}$ formation rates for mechanisms (i)(iii) and the partly preparation of the manuscript.

A. Jacob, S. F. Zhang, C. Müller, X. Ma, and A. B. Voitkiv, Phys. Rev. Research 2, 013105 (2020).

2. **Single ionization of an asymmetric diatomic system by relativistic charged projectiles**

This project considers the single electron emission from a weakly bound diatomic system by relativistic charged projectiles via direct and two-center impact ionization. The contribution of A. J. involves the analytical and numerical calculations of amplitudes and cross sections for direct and two-center impact ionization as well as the preparation of the manuscript.

A. Jacob, C. Müller, and A. B. Voitkiv, Phys. Rev. A 103, 042804 (2021).

3. **Formation of $\overline{\mathrm{H}}^+$ via electron-assisted three-body attachment of e^+ to $\overline{\mathrm{H}}$**

In this project, nonradiative attachment of a positron to antihydrogen in order to form the positive ion of antihydrogen $\overline{\mathrm{H}}^+$ via electron-assisted three-body attachment and positron-assisted three-body attachment is studied. The contribution of A. J. includes the analytical and numerical calculations of the $\overline{\mathrm{H}}^+$ formation rates for the three-body attachment mechanisms as well as the preparation of the manuscript.

A. Jacob, C. Müller, and A. B. Voitkiv, Phys. Rev. A 104, 032802 (2021).

4. **Radiation-field-driven ionization in laser-assisted slow atomic collisions**

This project considers the single electron emission in slow atomic collisions in the presence of a weak laser field via two-center resonant photoionization (2CPI) driven by the coupling of the colliding system to the radiation field. The contribution of A. J. involves the analytical calculation of the amplitude, cross section and reaction rate for 2CPI via the coupling to the radiation field as well as the preparation of the manuscript.

A. Jacob, C. Müller, and A. B. Voitkiv, arXiv:2208.09812 (2022).

Contents

Abbreviations[1]

2CDA	Two-center dileptonic attachment
2CPI	Two-center resonant photoionization
3BAe	Electron-assisted three-body attachment
3BAp	Positron-assisted three-body attachment
DPI	Direct photoionization
IAAD	Interatomic Auger decay
ICD	Interatomic Coulombic decay
LIRA	(Laser-)induced radiative attachment
SRA	Spontaneous radiative attachment

[1] In this thesis, we use abbreviations for the various (inter)atomic processes discussed. These are listed below:

List of Atomic Units[2]

Reduced Planck's constant	$\hbar = 1$ a. u. = 1.05×10^{-34} Js
Electron mass	$m_e = 1$ a. u. = 9.11×10^{-31} kg
Elementary charge	$e = 1$ a. u. = 1.60×10^{-19} C
Length	$a_0 = 1$ a. u. = 5.29×10^{-9} cm (Bohr radius for atomic hydrogen)
Velocity	$v_0 = 1$ a. u. = 2.19×10^{8} cm/s (electron velocity in 1st Bohr orbit of atomic hydrogen)
Energy	$E_h = 1$ a. u. = 4.36×10^{-18} J = 27.21 eV (Hartree energy which is twice the ionization potential of atomic hydrogen)
Time	$\tau_0 = 1$ a. u. = 2.42×10^{-17} s (time required for electron in 1st Bohr orbit of atomic hydrogen to travel one Bohr radius)
Angular frequency	$\omega_0 = 1$ a. u. = 6.58×10^{15} s^{-1} (angular frequency of electron in 1st Bohr orbit of atomic hydrogen) s^{-1}
Electric field strength	$F_0 = 1$ a. u. = 5.14×10^{9} V/cm (strength of Coulomb field experienced by electron in 1st Bohr orbit of atomic hydrogen)

[2] In this thesis, we use the system of atomic units (unless otherwise stated). In this system, the following applies:

List of Figures

Part I

Relativistic Effects in Interatomic Ionization Processes

Introduction and Preliminary Remarks

1

1.1 Historical Background and Motivation

The single ionization of an atom, i.e. the emission of a bound electron, by impact of ions or photoabsorption belong to the basic phenomena studied by atomic physics and can reveal insights into the structure of the atom as well as the interaction dynamics of the atom with its environment upon ionization. These well-known processes (for ion impact ionization see, e.g. [1–4] and references therein and for photoionization see, e.g. [5–8] and references therein) have been considered for a wide range of incident energies and, still today, there is great interest in further improving theoretical descriptions and experimental techniques.

The interaction between an atom and a (quantized) electromagnetic field can lead to various basic processes, e.g. resonant photon scattering, photoexcitation and photoionization, which have been studied for a long time. The famous photoelectric effect, in which electrons are emitted when electromagnetic radiation impinges on a (solid) material, was first experimentally observed by Hertz and Hallwachs in 1887/88 [9, 10] and later theoretically explained by Einstein in 1905 [11]. At the time, the understanding of the photoelectric effect was a very important step towards the development of quantum mechanics. Since then, there has been steady progress in experimental techniques especially due to more advanced light sources such as the optical laser, first completed in 1960 [12], which allowed the precise tuning of the incident light to atomic transitions. Besides, the development of synchrotron-based X-ray sources up to modern free electron lasers, which can produce ultra-short and -intense X-ray pulses, largely contributed to a more refined understanding of

---

Supplementary Information The online version contains supplementary material available at https://doi.org/10.1007/978-3-658-43891-3_1.

A. Jacob, *Relativistic Effects in Interatomic Ionization Processes and Formation of Antimatter Ions in Interatomic Attachment Reactions*,
https://doi.org/10.1007/978-3-658-43891-3_1

the (time-resolved) dynamics of photoionization mechanisms (see, e.g. [13] and references therein).

Ion-atom collisions can result in elastic scattering, charge exchange, impact excitation and impact ionization. A variety of such collision processes have been investigated for more than a century with one of the most prominent examples being the classical Rutherford scattering, in which alpha particles are elastically scattered on gold atoms. Experimentally observed by Geiger and Marsden in 1909 [14] and theoretically explained by Rutherford in 1911 [15], the Rutherford scattering process formed the early basis for today's picture of the atom. During the last decades the construction of accelerators that can produce energetic ion beams has seen a strong progress and enabled the investigation of fast ion-atom collisions up to highly relativistic impact energies, leading to a deeper understanding about the atomic structure and the dynamics involved in the collisions (see, e.g. [16, 17] and references therein).

In experiments on photoionization and ion impact ionization, the momenta of the product ions and/or ejected electrons have to be detected in order to rebuild the particles dynamics during the ionization process. Due to continuous development of more refined detectors, in modern experiments, one can obtain the particles momentum vectors by time of flight and position measurements using detector geometries with time- and position-sensitive detectors (see, e.g. [18, 19]).

There also exist more complex ionization mechanisms based on the decay of an autoionization state, which is an atomic bound state whose discrete energy level lies above the boundary of the continuous spectrum. An autoionization state arises, for instance, from the excitation of two atomic electrons whose total excitation energy exceeds the first ionization potential of the atom and it can decay, caused by electron correlations, where one electron makes a transition into a lower lying state and the second electron gets released by taking the energy excess. However, this is only one out of several types of autoionization states and associated ionization mechanisms (see, e.g. [20]).

Another type of autoionization state refers to the formation of an inner-shell vacancy by atomic excitation or ionization via photoabsorption or particle impact, where the excitation energy of the resulting atomic state is larger than its ionization potential. Such state is unstable and can decay through spontaneous radiative decay, in which an outer-lying electron fills the inner-shell vacancy and the energy release is carried away by spontaneous emission of a photon. However, the unstable state may also undergo radiationless decay, caused by electron correlations, where an outer-lying electron fills the inner-shell vacancy and the energy release is transferred to another outer-lying electron that, as a result, is ejected from the atom (see, e.g. [20]). The process involving the latter decay channel is called Auger process and the electron released upon decay is named Auger electron. The Auger process (or Auger-Meitner process) was independently discovered by Meitner in 1922 [21] and Auger in 1925 [22] and marked the beginning of investigations on atomic autoionization states.

Autoionization states can also exist in systems consisting of two atomic or molecular particles, where one of them is initially in an electronically excited state with the excitation energy exceeding the ionization potential of the other particle. There are various processes, driven by correlations between electrons located at two different atomic or molecular centers, which involve the relaxation of such nonlocal autoionization states. Penning ionization, originally considered by Penning in 1927 [23], typically occurs in collisions of an atom or molecule in a metastable excited electronic state (that is not allowed to decay by an optical dipole transition) with a second atom or molecule being in its electronic ground state. When the colliding partners approach each other very closely, their electronic orbitals overlap and radiationless relaxation of the initially excited particle with simultaneous ionization of the other particle (mainly) proceeds via charge transfer. This ionization process is of very short range and its efficiency decreases exponentially with increasing interatomic/intermolecular distance. Motivated by experimental studies on de-excitation processes in metallic compounds by Gallon and Matthew in 1970 [24] as well as Lord and Gallon in 1973 [25], the radiationless relaxation of an optically excited atom by energy transfer to a neighboring ground-state atom, resulting in its ionization, was theoretically considered by Matthew and Komninos in 1975 [26] and called interatomic Auger decay (IAAD). Here, the excited atom initially has an inner-shell vacancy which is filled by an outer-lying electron but unlike in the intraatomic Auger process (that was discussed further above) the energy release is transferred via a (long-range) dipole-dipole interaction to a neighboring ground-state atom and the latter is ionized. It was shown in [26] that IAAD is very efficient and can even dominate over the competing intraatomic Auger process. The mechanism of IAAD proceeding in a system of two atomic species where relaxation of the excited particle via the intraatomic Auger process is energetically forbidden was calculated by Cederbaum et. al. in 1997 [27] and termed there interatomic Coulombic decay (ICD). It was experimentally observed by Marburger et. al. in 2003 [28] and Jahnke et. al. in 2004 [29]. The ICD mechanism can be extremely efficient, strongly outperforming the (direct) spontaneous radiative decay of the initially electronically excited atom, for a large range of interatomic separations. We mention that the term ICD is now often understood more generally to refer to all long-range interatomic (or intermolecular) radiationless relaxation mechanisms proceeding at large distances via the dipole-dipole interaction.

A well-known excitation mechanism caused by long-range electron correlations between neighboring molecules is Förster resonance energy transfer which was described by Förster in 1948 [30]. This energy transfer mechanism can occur between chromophores (light-sensitive molecules) in biological systems [31], where an initially electronically excited chromophore may transfer energy to another chromophore via a (long-range) dipole-dipole interaction such that the latter makes a transition from its electronic ground state into an electronically excited bound state.

In general, the effectiveness of electron correlations between neighboring atoms and/or molecules makes the theoretical and experimental investigation of processes based on such long-range correlations particularly interesting. Therefore, within the last two decades, several more nonlocal electron correlation phenomena were considered including electron transfer mediated decay [32], interatomic Coulombic electron capture in weakly bound systems [33] and in atomic collisions [34], collective autoionization/ICD (see, e.g. [35–38] and references therein), two-center resonance scattering [39] as well as two-center dielectronic recombination in weakly bound systems [40] and in atomic collisions [41]. (The latter process will be discussed in detail in Part II of this thesis in the context of the formation of positive ions of antihydrogen.)

Besides, there exist two additional mechanisms for single ionization of diatomic systems, caused by interatomic electron correlations, that are crucial for the present study. They will be discussed in the following Section.

1.2 Overview of Direct and Two-Center Impact and Photoionization Mechanisms

One of the interatomic ionization processes which will be considered in this thesis is termed two-center impact ionization,

$$\text{(i)} \quad A + B + P \;\rightarrow\; A + B^* + P \;\rightarrow\; A^+ + \mathrm{e}^- + B + P.$$

It involves the excitation of a dipole-allowed transition in an atomic species B by impact of a charged projectile P (e.g. an ion or electron) with subsequent relaxation via ICD that means the radiationless decay of the excited state of B by transmitting the de-excitation energy – due to interatomic electron correlations – to a neighboring atom A which, as a consequence, is ionized. This process was considered (using an instantaneous dipole-dipole interaction) for nonrelativistic electron impact in [42], where it was concluded that the two-center mechanism can substantially enhance the total electron emission from the $A - B$ system. In this work, based on the results of [43], two-center impact ionization will be further developed by considering the impact of relativistic bare ions and by taking into account that the interatomic interaction propagates with a finite velocity, resulting in the retardation effect.

The other interatomic process is called two-center resonant photoionization (2CPI),

$$\text{(ii)} \quad A + B + N\hbar\omega \rightarrow A + B^* + (N-1)\hbar\omega \rightarrow A^+ + e^- + B + (N-1)\hbar\omega,$$

in which the diatomic system, consisting of atoms A and B, is exposed to a weak laser field with frequency ω that is resonantly tuned to a dipole-allowed transition in B. Then, excitation of B occurs via (resonant) photoabsorption with consequent relaxation by ICD, i.e. via the transfer of the excitation energy to A, induced by interatomic electron correlations, causing its ionization. 2CPI was originally considered for a weakly bound system in [44, 45], where it was shown that this ionization mechanism can strongly outperform the direct photoionization of A by the laser field. Its very high efficiency was experimentally confirmed in experiments on photoionization of Ne–He dimers [46, 47] and Ar-Ne clusters [48]. 2CPI was also studied theoretically in slow atomic collisions [49], where the interaction between the colliding atoms was regarded as instantaneous, being transmitted by virtual (off-shell) photons only. In this work, following the consideration in [50], we extend the theory of collisional 2CPI to a more complete treatment, in which the collisional interaction is described fully relativistically, accounting for the retardation effect, that opens the possibility to transmit the interaction by real (on-shell) photons.

It is of general interest to discuss the relative effectiveness of the two-center ionization channels (i) and (ii) compared with the corresponding direct ionization by ion impact,

$$\text{(iii)} \quad A + P \rightarrow A^+ + e^- + P,$$

and by photoabsorption in the presence of a laser field,

$$\text{(iv)} \quad A + N\hbar\omega \rightarrow A^+ + e^- + (N-1)\hbar\omega.$$

In reaction (iii), atom A is ionized as a direct result of the collisional interaction with the incident ion P, while in reaction (iv) the ionization of A is a direct consequence of the absorption of a photon with frequency ω from a weak laser field.

Indeed, we will see that the interatomic ionization mechanisms of (i) two-center ion impact ionization and (ii) two-center resonant photoionization (in atomic collisions) can outperform the corresponding local ionization mechanisms of (iii) direct ion impact ionization and (iv) direct photoionization, respectively.

Atomic units (see overview on p. xxiii) are used throughout if not stated otherwise.

1.3 Relativistic Effects in Two-Center Impact and Photoionization

In general, there exist three types of relativistic effects that may be relevant to our consideration of ionization of diatomic systems. The first and second type are based on a large electron orbiting and projectile impact velocity, respectively, and the third one is caused by a large interatomic distance.

The first type of relativistic effect is due to a large orbiting velocity v_e of atomic electrons, approaching the speed of light c ($c \approx 137$ a.u.). In such case, the relativistic motion of electrons involved in atomic excitation and ionization processes has to be described by the Dirac equation. However, in this thesis, we consider only relatively light atomic species with atomic number $Z \ll c$ for which the description of the electronic motion in atomic excitation and ionization processes by the nonrelativistic Schrödinger equation is regarded as valid (see, e.g. [51]). It is worth mentioning that for ionization by impact of charged projectiles the vast majority of electrons emitted from light atomic targets have nonrelativistic velocities in the target frame even for extremely relativistic impact energies [51].

The second type of relativistic effect results from a large projectile impact velocity v that approaches the speed of light c. It will be investigated in the context of impact ionization of a weakly bound diatomic system by relativistic bare ions in Chapter 2.

To get an idea about the general effect resulting from relativistic collision velocities, let us suppose that a pointlike positively charged projectile (representing e.g. a bare ion) is incident on a target atom whose nucleus is located at the origin of the coordinate system. The projectile moves along a classical straight-line trajectory $\boldsymbol{R}(t) = \boldsymbol{b} + \boldsymbol{v}t$, where $\boldsymbol{v} = (0, 0, v)$ is its constant velocity and $\boldsymbol{b} = (b, 0, 0)$ is the impact parameter for the collision. In such situation, the scalar potential $\phi(\boldsymbol{r}, t)$ and vector potential $\boldsymbol{A}(\boldsymbol{r}, t)$ produced by the projectile in the restframe of the target at the point of observation $\boldsymbol{r} = (x, y, z)$ at the time t are given by the Liénard-Wiechert potentials in the forms [2]

$$\phi(\boldsymbol{r}, t) = \frac{\gamma Z_P}{\sqrt{(x-b)^2 + y^2 + \gamma^2 (z - vt)^2}}, \quad \boldsymbol{A}(\boldsymbol{r}, t) = \frac{\boldsymbol{v}}{c}\phi(\boldsymbol{r}, t). \tag{1.1}$$

Here, $\gamma = 1/\sqrt{1-\beta^2}$ is the Lorentz factor with the reduced velocity $\beta = v/c$ and Z_P is the projectile charge. The potentials in (1.1) satisfy the Lorenz gauge condition $\frac{1}{c}\frac{\partial\phi}{\partial t} + \nabla \cdot \boldsymbol{A} = 0$.

Following the consideration in [2], the corresponding electric field at the observation point located at the position of the target nucleus is directed radially from the present position of the projectile to the observation point and is obtained to be

$$\boldsymbol{E} = \frac{-Z_P \boldsymbol{R}}{\gamma^2 R^3 (1-\beta^2 \sin^2\theta)^{3/2}}, \tag{1.2}$$

where θ is the angle between the vector $-\boldsymbol{R}$ and the z-axis. In the longitudinal direction with respect to the projectile motion ($\theta = 0$ or $\theta = \pi$) the field strength becomes

$$E_{\parallel} = \frac{Z_P}{R^2}\frac{1}{\gamma^2}. \tag{1.3}$$

It is decreased by a factor γ^{-2} compared with the field strength for a point charge at rest. Moreover, in the transverse direction with respect to the projectile motion ($\theta = \pi/2$) the field strength reads

$$E_{\perp} = \frac{Z_P}{R^2}\gamma. \tag{1.4}$$

It is increased by a factor γ as compared to the field strength for a point charge at rest. One can think of this as a flattening of the electric field of a moving charge into a disk-like shape in the direction of motion that arises from the Lorentz contraction of electromagnetic fields. For instance, the flattening of the electric field for $\gamma = 2$ (corresponding to $v \approx 0.87c$) is already quite pronounced with a decrease in the longitudinal field strength by a factor of 4 and an increase in the transverse field strength by a factor of 2.

The third type of relativistic effect which may occur in a system of two interacting atoms is the so-called retardation effect. It accounts for the finite propagation time of the electromagnetic field transmitting the interaction with the velocity c and becomes relevant at large interatomic distances. The retardation effect will be investigated in the context of ion impact ionization of a weakly bound diatomic system in Chapter 2 as well as in the context of two-center resonant photoionization in slow atomic collisions in Chapter 3.

In order to estimate the importance of the retardation effect, we may compare the propagation time $T = R/c$ that is necessary for the electromagnetic field to propagate the distance R between the atoms and the electronic transition time τ. In case when $T \ll \tau$, the field propagates essentially instantaneously and the retardation effect is expected to have no substantial impact on the interatomic interaction. On the other hand, if $T \gg \tau$, the comparatively large propagation time of the field results in a retardation effect which strongly influences the interaction between the atoms dramatically increasing its effective range. The latter condition can be rewritten as $R \gg \tau c$, indicating that the interatomic distance R must be sufficiently large for the retardation effect to be significant. For instance, assuming a typical electronic transition time $\tau \sim 1$ a.u., we can expect that the retardation effect is important when $R \gg 10^2$ a.u.

In particular, let us consider a nonlocal energy transfer in a diatomic system consisting of two atoms A and B with a relatively large interatomic distance R, where the transfer of energy is caused by the long-range interatomic electron correlations between an electron in A and another electron in B. Then, the interatomic interaction can be described in very good approximation by the dipole-dipole interaction (see Section 2.1.1 and Appendix 9.1 in the Electronic Supplementary Material for a detailed derivation)

$$\hat{V}_{AB} = e^{i\frac{R}{\tau c}}\left[\left(\boldsymbol{r}\cdot\boldsymbol{\xi} - \frac{3(\boldsymbol{r}\cdot\boldsymbol{R})(\boldsymbol{\xi}\cdot\boldsymbol{R})}{R^2}\right)\frac{1-i\frac{R}{\tau c}}{R^3} - \left(\boldsymbol{r}\cdot\boldsymbol{\xi} - \frac{(\boldsymbol{r}\cdot\boldsymbol{R})(\boldsymbol{\xi}\cdot\boldsymbol{R})}{R^2}\right)\frac{\left(\frac{1}{\tau c}\right)^2}{R}\right] \tag{1.5}$$

with $\boldsymbol{R}$ the interatomic distance vector, which is constant for a bound diatomic system and depends on the time t in the case of atomic collisions. Further, $\boldsymbol{r}$ ($\boldsymbol{\xi}$) is the coordinate of the partaking electron in atom A (B) with respect to the nucleus of A (B).

The dependence of (1.5) on the interatomic distance R in leading order is given by $\hat{V}_{AB} \sim R^{-1}$. In the limit of comparatively small propagation times $T \ll \tau$, corresponding to comparatively small interatomic distances $R \ll \tau c$, the interaction (1.5) takes the familiar form (see, e.g. [52]) of the instantaneous interaction between two electric dipoles

$$\hat{V}_{AB} = \left(\boldsymbol{r}\cdot\boldsymbol{\xi} - \frac{3(\boldsymbol{r}\cdot\boldsymbol{R})(\boldsymbol{\xi}\cdot\boldsymbol{R})}{R^2}\right)\frac{1}{R^3}, \tag{1.6}$$

which scales with the interatomic distance as R^{-3}.

Note that the general form of the dipole-dipole interaction in (1.5) takes into account the relativistic retardation effect resulting from the finite propagation of the electromagnetic interaction. It can be concluded from the form of (1.5) that the retardation effect starts to become important for interatomic distances $R \gtrsim \tau c$, changing the R-dependence of $\hat{V}_{AB}$ from $\sim R^{-3}$ at $R \ll \tau c$ to $\sim R^{-1}$ at $R \gg \tau c$ and therefore tremendously increasing the effective range of the interatomic interaction.

One of the main goals of this study is to investigate the influence of the relativistic effects discussed above on interatomic ionization processes.

Concerning two-center ion impact ionization of a weakly bound diatomic system, it will be shown that the influence of relativistic effects, resulting from a high ion impact velocity, on the angular distribution of emitted electrons can be quite strong already at rather low Lorentz factors of $\gamma = 1/\sqrt{1 - v^2/c^2} \approx 2$. On the other hand, these effects may only have a substantial impact on the energy distribution and the total cross section for $\gamma \gg 1$. In addition, we will see that the relativistic retardation effect, taking into account the finite propagation of the electromagnetic interaction between the atoms, has essentially no influence on the two-center ionization even for rather large diatomic systems such as the ^{7}Li–He dimer whose mean size is ≈ 53 a.u.

However, this dramatically changes when considering atomic collisions instead of a weakly bound system. As an example, we take the process of two-center resonant photoionization in atomic collisions, where the collision velocity v shall be much smaller than the typical orbiting velocities v_e of the participating electrons. We will see that in collisions the retardation effect, accounting for the finite propagation of the interaction between the colliding atoms, leads to an efficient coupling of the diatomic system to the quantum radiation field. This enables the interaction to proceed via the exchange of an on-shell photon, thus dramatically increasing its effective range, which may profoundly modify the two-center photoionization process and strongly enhance its reaction rate.

Part I of this thesis is essentially organized as follows. In Chapter 2, we consider the theoretical framework for the single ionization of a weakly bound diatomic system by relativistic charged projectiles and obtain formulas for differential and total cross sections of direct and two-center ion impact ionization, respectively. Afterwards, numerical results are illustrated and extensively discussed. Finally, we draw some main conclusions. Chapter 3 is dedicated to the theory of radiation-field-driven ionization in laser-assisted slow atomic collisions, where we derive simple formulas for the cross section and reaction rate of the two-center photoionization process. We then present numerical results, discuss them in detail and summarize our main findings.

Ionization of a Weakly Bound Diatomic System by Relativistic Charged Projectiles

2

This chapter provides a detailed insight into the theoretical treatment of the single electron emission from a weakly bound diatomic system by relativistic charged projectiles via direct and two-center impact ionization. We derive the angular and energy distributions as well as the total cross section for these ionization mechanisms. Based on their numerical results we discuss, in particular, the influence of the relativistic effects, caused by a large projectile velocity and a large size of the diatomic system, on the two-center ionization channel. Besides, we also consider the relative effectiveness of two-center impact ionization compared with direct impact ionization. The following chapter is mainly based on results published initially in Ref. [43].

2.1 Theoretical Consideration

2.1.1 General Approach

Let us consider a weakly bound diatomic system consisting of two atomic species A and B which are in their electronic ground states with energies ε_g and ϵ_g, respectively. We assume that there exist an excited state with energy ϵ_e in atom B which can be populated via a dipole-allowed transition from its ground state, where the corresponding transition energy $\omega_B = \epsilon_e - \epsilon_g$ shall be larger than the ionization

Supplementary Information The online version contains supplementary material available at https://doi.org/10.1007/978-3-658-43891-3_2.

A. Jacob, *Relativistic Effects in Interatomic Ionization Processes and Formation of Antimatter Ions in Interatomic Attachment Reactions*,
https://doi.org/10.1007/978-3-658-43891-3_2

potential I_A ($= -\varepsilon_g$) of atom A. Further, it is supposed that the interatomic distance R between the nuclei of A and B is much larger than the typical atomic size so that the electronic orbitals of atoms A and B essentially do not overlap and the interaction between them is relatively weak. Consequently, the electronic structure of the $A - B$ system may be approximated by that of two individual non-interacting atomic species A and B.

Now, let us envisage a situation in which the $A-B$ system is bombarded by a bare ion P with charge Z_P and relative (with respect to the diatomic system) velocity v. In the present treatment, we shall suppose the condition $Z_P/v \ll 1$, which means that we only consider relatively light projectiles having comparatively low charges. Then, the electric field of the projectile will just represent a weak perturbation for the $A - B$ system and we may treat the interaction between this system and the projectile by using the first order of time dependent perturbation theory. Moreover, the above condition also implies that the (total) ionization of the $A - B$ system will be largely dominated by ionization processes leading to single electron emission.

Single ionization of the $A - B$ system in collisions with charged projectiles may occur via two direct (one-step) and one indirect (two-step) ionization processes. The two direct processes are the direct impact ionization of either atom A or B due to the collisional interaction of the incident ion P with A or B. The indirect process is two-center impact ionization which involves both atomic species A and B. In the first step of this process, the collisional interaction between the incident ion P and atom B causes a dipole-allowed transition from the ground state of B with energy ϵ_g into its excited state with energy ϵ_e. In the second step, atom B radiationlessly decays into its initial ground state where the released energy is transferred – via long-range interatomic electron correlations – to atom A which consequently undergoes a transition from its ground state with energy ε_g into a continuum state with energy ε_k. A scheme of two-center impact ionization is illustrated in Fig. 2.1 (a). Note that due to the condition $\epsilon_e - \epsilon_g > I_A$ we just have to deal with two-center impact ionization of atom A and not of atom B.

In contrast to direct impact ionization, two-center ionization is a resonant process which only proceeds efficiently within a very narrow interval of electron emission energies centered at the resonance energy $\varepsilon_{k_r} = (\epsilon_e - \epsilon_g) + \varepsilon_g = \omega_B - I_A$. However, as we will see, the two-center ionization is so tremendously strong on and close to the resonance that, despite the energy range affected by the two-center channel being very small, its presence can even have a noticeable impact on the total ionization cross section.

In our present treatment of all ionization channels, we use the single-electron approximation and thus only consider one active electron in each atom A and B. Further, our description of collisions between the projectile P and the $A - B$ system

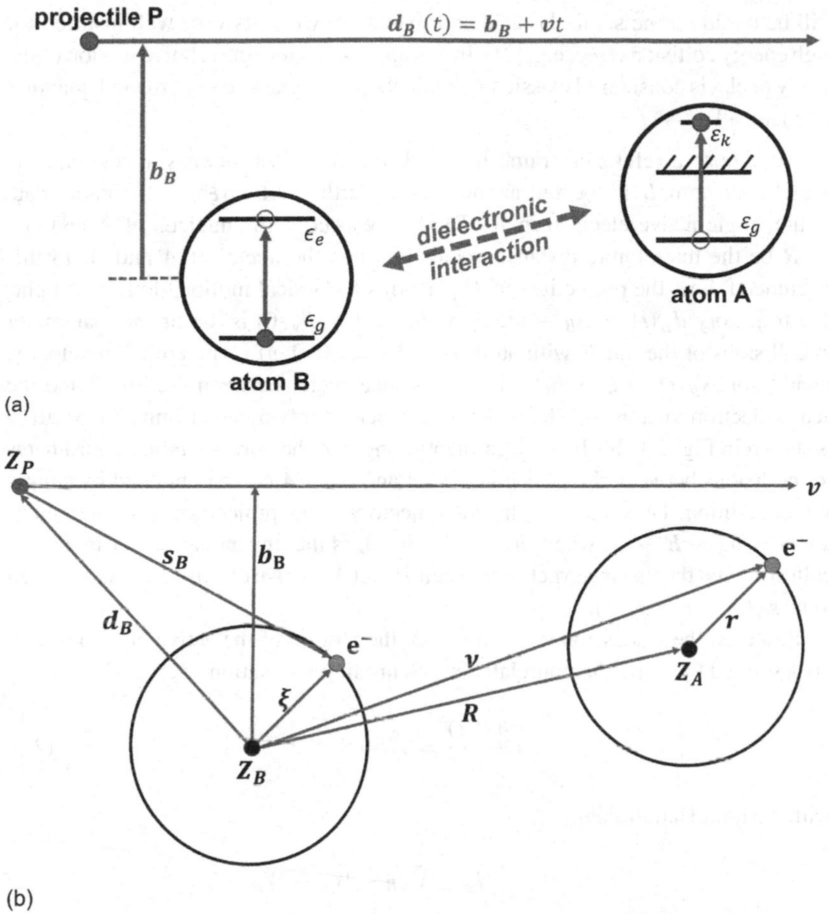

Fig. 2.1 (a) Scheme of two-center ion impact ionization. (b) Schematic representation of space coordinates characterizing the collision. (This figure was originally published in Ref. [43])

will be based on the semi-classical approximation which is very well justified for high energy collisions (see, e.g. [2]). In this approximation, the relative motion of the heavy nuclei is considered classically while the active electrons are treated quantum mechanically.

We choose a reference frame in which the $A - B$ system is at rest and the nucleus of atom B is located at the origin. Further, let $\boldsymbol{r}$ ($\boldsymbol{\xi}$) be the coordinate of the single active electron in A (B) with respect to the nucleus of A (B) and let $\boldsymbol{R}$ be the interatomic distance vector between the nuclei of A and B. In this reference frame, the projectile ion P performs a classical motion along a straight-line trajectory $\boldsymbol{d}_B(t) = \boldsymbol{b}_B + \boldsymbol{v}t$. Here, $\boldsymbol{b}_B = (b_x, b_y, 0)$ is the impact parameter in collisions of the ion P with atom B and $\boldsymbol{v} = (0, 0, v)$ is the collision velocity. In addition, $\boldsymbol{s}_B(t) = \boldsymbol{\xi} - \boldsymbol{d}_B(t)$ is the distance vector between the ion P and the active electron in atom B. The collision geometry for two-center impact ionization is shown in Fig. 2.1 (b). It is worth mentioning that the corresponding coordinates for collisions between the projectile ion P and atom A can be obtained by simple vector addition. Thus, the straight-line trajectory of the projectile can be written as $\boldsymbol{d}_A(t) = \boldsymbol{b}_A - \boldsymbol{R} + \boldsymbol{v}t$, where $\boldsymbol{b}_A = (\tilde{b}_x, \tilde{b}_y, 0)$ is the impact parameter in $P - A$ collisions and the distance vector between P and the active electron in A is obtained to be $\boldsymbol{s}_A(t) = \boldsymbol{r} - \boldsymbol{d}_A(t)$.

Based on the discussion in Section 1.3, the motion of the active electrons may be described by using the nonrelativistic Schrödinger equation

$$i\frac{\partial\Psi(t)}{\partial t} = \hat{H}\Psi(t) \tag{2.1}$$

with the total Hamiltonian

$$\hat{H} = \hat{H}_A + \hat{H}_B + \hat{V}_{AB} + \hat{W}_A + \hat{W}_B. \tag{2.2}$$

In (2.2), $\hat{H}_A$ ($\hat{H}_B$) is the Hamiltonian of the free non-interacting atom A (B), $\hat{V}_{AB}$ is the interaction between A and B, and $\hat{W}_A$ ($\hat{W}_B$) is the interaction between the projectile ion P and atom A (B).

The free Hamiltonians $\hat{H}_A$ and $\hat{H}_B$ of atoms A and B, respectively, are given by

$$\begin{aligned} \hat{H}_A &= \frac{(\hat{\boldsymbol{p}}_r)^2}{2} - \frac{Z_A}{r}, \\ \hat{H}_B &= \frac{(\hat{\boldsymbol{p}}_\xi)^2}{2} - \frac{Z_B}{\xi}. \end{aligned} \tag{2.3}$$

Here, $\hat{\boldsymbol{p}}_r$ ($\hat{\boldsymbol{p}}_\xi$) is the momentum operator for the active electron in A (B) with respect to the nucleus of A (B) and Z_A (Z_B) is the effective nuclear charge of atom A (B).

The interaction $\hat{V}_{AB}$ between atoms A and B at relatively large interatomic distances R is primarily of the dipole-dipole type and can be derived by considering the coupling $j^A_\mu A^\mu_B$ between the transition four-current j^A_μ of the active electron in A and the four-potential A^μ_B of the field created by the other active electron in B (or vice versa). Its detailed derivation can be found in Appendix 9.1 in the Electronic Supplementary Material and results in the expression

$$\hat{V}_{AB} = e^{iR\frac{\omega_B}{c}}\left[\left(\boldsymbol{r}\cdot\boldsymbol{\xi} - \frac{3(\boldsymbol{r}\cdot\boldsymbol{R})(\boldsymbol{\xi}\cdot\boldsymbol{R})}{R^2}\right)\frac{1-iR\frac{\omega_B}{c}}{R^3}\right.$$
$$\left. - \left(\boldsymbol{r}\cdot\boldsymbol{\xi} - \frac{(\boldsymbol{r}\cdot\boldsymbol{R})(\boldsymbol{\xi}\cdot\boldsymbol{R})}{R^2}\right)\frac{\left(\frac{\omega_B}{c}\right)^2}{R}\right]. \tag{2.4}$$

Note that (2.4) yields the form in (1.5) by applying $\tau = 1/\omega_B$. The interaction in (2.4) incorporates the relativistic retardation effect, accounting for the finite propagation of the electromagnetic interaction, and its limit for comparatively small interatomic distances, $R \ll c/\omega_B$, is the instantaneous interaction between two electric dipoles given by (1.6). Further details can be found in Section 1.3.

Now, we consider the interactions $\hat{W}_A$ and $\hat{W}_B$ between the projectile ion P and atoms A and B, respectively. Keeping in mind that we suppose a nonrelativistic electron motion these interactions read

$$\hat{W}_A = \frac{1}{2c}\left[\hat{\boldsymbol{p}}_r\cdot\boldsymbol{A}_A + \boldsymbol{A}_A\cdot\hat{\boldsymbol{p}}_r\right] - \phi_A + \frac{1}{2c^2}\boldsymbol{A}_A^2 + \frac{1}{2c}\boldsymbol{\sigma}\cdot\boldsymbol{B}_A,$$
$$\hat{W}_B = \frac{1}{2c}\left[\hat{\boldsymbol{p}}_\xi\cdot\boldsymbol{A}_B + \boldsymbol{A}_B\cdot\hat{\boldsymbol{p}}_\xi\right] - \phi_B + \frac{1}{2c^2}\boldsymbol{A}_B^2 + \frac{1}{2c}\boldsymbol{\sigma}\cdot\boldsymbol{B}_B \tag{2.5}$$

with ϕ_A (ϕ_B) and $\boldsymbol{A}_A$ ($\boldsymbol{A}_B$) the scalar and vector potential, respectively, which determine the electromagnetic field of the projectile ion P acting on the active electron in A (B). These potentials can be described by the Liénard-Wiechert potentials whose forms are similar to those in (1.1) but are adapted to the present collision geometry. They can be written as

$$\phi_A = \frac{\gamma Z_P}{|\boldsymbol{s}'_A(t)|}, \quad \boldsymbol{A}_A = \frac{\boldsymbol{v}}{c}\phi_A,$$
$$\phi_B = \frac{\gamma Z_P}{|\boldsymbol{s}'_B(t)|}, \quad \boldsymbol{A}_B = \frac{\boldsymbol{v}}{c}\phi_B, \tag{2.6}$$

where $\gamma = 1/\sqrt{1-\beta^2}$ is the Lorentz factor with the reduced velocity $\beta = v/c$ and

$$s'_A(t) = \big(\boldsymbol{r}_\perp - \boldsymbol{b}_A - \boldsymbol{R}_\perp, \gamma(r_\parallel - R_\parallel - vt)\big),$$
$$s'_B(t) = \big(\boldsymbol{\xi}_\perp - \boldsymbol{b}_B, \gamma(\xi_\parallel - vt)\big). \tag{2.7}$$

Here, $\boldsymbol{r}_\perp$ ($r_\parallel$), $\boldsymbol{\xi}_\perp$ ($\xi_\parallel$) and $\boldsymbol{R}_\perp$ ($R_\parallel$) are the transverse (longitudinal) parts of the coordinates $\boldsymbol{r}$, $\boldsymbol{\xi}$ and $\boldsymbol{R}$, respectively, which are perpendicular (parallel) to the collision velocity $\boldsymbol{v}$. It is worth mentioning that the potentials in (2.6) satisfy the Lorenz condition $\partial_\mu A_j^\mu = 0$ for the four-potential $A_j^\mu = (\phi_j, \boldsymbol{A}_j)$ with $j = A, B$.

Further, in (2.5), $\boldsymbol{\sigma}$ are the Pauli matrices and $\boldsymbol{B}_A$ ($\boldsymbol{B}_B$) is the magnetic field of the projectile ion P acting on the active electron in atom A (B). Since it is known (see, e.g. [53]) from the theory of direct impact ionization that spin effects are negligible for light atomic targets, we drop the corresponding interaction term $\frac{1}{2c}\boldsymbol{\sigma} \cdot \boldsymbol{B}_j$ ($j = A, B$) in our consideration.

The interaction term $\frac{1}{2c^2}\boldsymbol{A}_j^2$ ($j = A, B$) in (2.5) requires special care. In particular, it was shown in [51] that this term has to be omitted in the Schrödinger equation in order to obtain a self-consistent first order treatment. Consequently, we drop this term in our consideration as well.

The initial state Ψ_{gg} of the $A - B$ system is given by

$$\Psi_{gg}(\boldsymbol{\xi}, \boldsymbol{\nu}, t) = \phi_g(\boldsymbol{\nu} - \boldsymbol{R})e^{-i\varepsilon_g t}\chi_g(\boldsymbol{\xi})e^{-i\epsilon_g t} \tag{2.8}$$

with ϕ_g (χ_g) the ground state of atom A (B) and $\boldsymbol{\nu} = \boldsymbol{R} + \boldsymbol{r}$. Concerning the direct impact ionization of atom A, the final state $\Psi_{\boldsymbol{k}g}$ is determined by

$$\Psi_{\boldsymbol{k}g}(\boldsymbol{\xi}, \boldsymbol{\nu}, t) = \phi_{\boldsymbol{k}}(\boldsymbol{\nu} - \boldsymbol{R})e^{-i\varepsilon_k t}\chi_g(\boldsymbol{\xi})e^{-i\epsilon_g t}. \tag{2.9}$$

In (2.9), $\phi_{\boldsymbol{k}}$ is the continuum state of the electron emitted from A with an asymptotic momentum $\boldsymbol{k}$ and energy $\varepsilon_k = \boldsymbol{k}^2/2$. Accordingly, the final state $\Psi_{g\boldsymbol{\kappa}}$ for the direct impact ionization of atom B reads

$$\Psi_{g\boldsymbol{\kappa}}(\boldsymbol{\xi}, \boldsymbol{\nu}, t) = \phi_g(\boldsymbol{\nu} - \boldsymbol{R})e^{-i\varepsilon_g t}\chi_{\boldsymbol{\kappa}}(\boldsymbol{\xi})e^{-i\epsilon_\kappa t}. \tag{2.10}$$

Here, $\chi_{\boldsymbol{\kappa}}$ is the continuum state of the electron ejected from B having an asymptotic momentum $\boldsymbol{\kappa}$ and energy $\epsilon_\kappa = \boldsymbol{\kappa}^2/2$. Considering two-center impact ionization of atom A, in addition to the initial state (2.8) and the final state (2.9), we also have to take into account the intermediate state(s)

$$\Psi_{ge}(\boldsymbol{\xi}, \boldsymbol{\nu}, t) = \phi_g(\boldsymbol{\nu} - \boldsymbol{R})e^{-i\varepsilon_g t}\chi_e(\boldsymbol{\xi})e^{-i\epsilon_e t}, \tag{2.11}$$

where χ_e is the excited state of B.

The channels for two-center and direct impact ionization of atom A result in the same final state (2.9) of the $A - B$ system and thus they interfere with each other. Consequently, the transition amplitude for the ionization of A is composed of the two-center amplitude $\mathcal{S}_{2C}(\boldsymbol{b}_B)$ and direct amplitude $\mathcal{S}_D^A(\boldsymbol{b}_A)$ according to

$$\mathcal{S}_{2C+D}(\boldsymbol{b}_B, \boldsymbol{b}_A) = \mathcal{S}_{2C}(\boldsymbol{b}_B) + \mathcal{S}_D^A(\boldsymbol{b}_A). \tag{2.12}$$

We remark that the channel for direct impact ionization of atom B leads to a different final state (2.10) and therefore does not interfere with the other two ionization channels.

However, it is worth mentioning that the impact ionization of atom B may trigger subsequent radiative capture of an electron from atom A by the residual ion B^+, resulting in the same $A^+ - B$ system as the direct and two-center impact ionization of A. In particular, the B^+ ion formed via impact ionization of B will polarize atom A leading to the appearance of an attractive force between B^+ and A. In case the two atomic centers approach sufficiently close each other, an outer shell electron of atom A may be captured by the B^+ ion accompanied by the emission of a photon that results in a diatomic system consisting of the neutral atom B in its ground state and the ion A^+. Since the above described process involves the interaction between two atomic centers, namely the B^+ ion and atom A, it may also be seen as a kind of two-center ionization which might significantly increase the number of collision events leading to the $A^+ - B$ system with an associated decrease in the number of collision events which result in the $A - B^+$ system. However, in contrast to two-center impact ionization, the process involving radiative electron capture is not resonant, having a shape of the electron emission spectrum similar to that of the direct impact ionization of atom B, and we will not consider it in this work.

2.1.2 Amplitude for Two-Center Impact Ionization of Atom *A*

The transition amplitude $\mathcal{S}_{2C}(\boldsymbol{b}_B)$ for two-center impact ionization of atom A in (2.12) is described by using the second order of time dependent perturbation theory in which both the interaction $\hat{W}_B$ between the projectile ion P and atom B and the interatomic interaction $\hat{V}_{AB}$ between atoms A and B are accounted for and it can be written as

$$\mathcal{S}_{2C}(\boldsymbol{b}_B) = \sum_{\Delta m=-1}^{1} \mathcal{S}_{2C}^{\Delta m}(\boldsymbol{b}_B). \tag{2.13}$$

In (2.13), $\Delta m \in \{0, \pm 1\}$ indicates the change in the magnetic quantum number for the dipole-allowed excitation transition in B and

$$\mathcal{S}_{2C}^{\Delta m}(\boldsymbol{b}_B) = \frac{1}{i^2} \int_{-\infty}^{\infty} dt\, \mathcal{M}_2^{\Delta m}(t) \int_{-\infty}^{t} dt'\, \mathcal{M}_1^{\Delta m}(\boldsymbol{b}_B, t') \tag{2.14}$$

with $\mathcal{M}_1^{\Delta m}(\boldsymbol{b}_B, t') = \langle \Psi_{ge} | \hat{W}_B(\boldsymbol{b}_B, t') | \Psi_{gg} \rangle$ and $\mathcal{M}_2^{\Delta m}(t) = \langle \Psi_{\boldsymbol{k}g} | \hat{V}_{AB} | \Psi_{ge} \rangle$. Inserting the states from (2.8), (2.9) and (2.11) into (2.14) and taking advantage of the orthonormalization condition $\langle \phi_g | \phi_g \rangle = 1$ provides

$$\mathcal{S}_{2C}^{\Delta m}(\boldsymbol{b}_B) = \frac{1}{i^2} \int_{-\infty}^{\infty} dt\, \mathcal{M}_{AB}^{\Delta m} e^{i(\omega_A - \omega_B)t} \int_{-\infty}^{t} dt'\, \mathcal{M}_B^{\Delta m}(\boldsymbol{b}_B, t') e^{i\omega_B t'}, \tag{2.15}$$

where

$$\mathcal{M}_{AB}^{\Delta m} = \langle \phi_{\boldsymbol{k}} \chi_g | \hat{V}_{AB} | \phi_g \chi_e \rangle \tag{2.16}$$

is the interatomic matrix element describing the de-excitation in atom B and the ionization of atom A and

$$\mathcal{M}_B^{\Delta m} = \langle \chi_e | \hat{W}_B | \chi_g \rangle \tag{2.17}$$

is the matrix element for the impact excitation of B. Further, $\omega_A = \varepsilon_k - \varepsilon_g$ is the transition energy in A.

By defining $F(t) = \int_{-\infty}^{t} dt'\, \mathcal{M}_B^{\Delta m}(\boldsymbol{b}_B, t') e^{i\omega_B t'}$, the transition amplitude in (2.15) yields

$$\mathcal{S}_{2C}^{\Delta m}(\boldsymbol{b}_B) = \frac{1}{i^2} \int_{-\infty}^{\infty} dt\, \mathcal{M}_{AB}^{\Delta m} e^{i(\omega_A - \omega_B)t} F(t). \tag{2.18}$$

Taking into account that the interatomic matrix element $\mathcal{M}_{AB}^{\Delta m}$ is constant for finite t and vanishes at the boundaries $t = \pm\infty$, which corresponds to the assumption that the interaction between atoms A and B is adiabatically switched on and off at $t \to -\infty$ and $t \to +\infty$, respectively, integration by parts in (2.18) results in

$$S_{2C}^{\Delta m}(\boldsymbol{b}_B) = \frac{-i\mathcal{M}_{AB}^{\Delta m}}{\omega_A - \omega_B + i\gamma/2} \int_{-\infty}^{\infty} dt\, \mathcal{M}_B^{\Delta m}(\boldsymbol{b}_B, t) e^{i\omega_A t}. \tag{2.19}$$

Here, the appearance of γ ($\gamma \to 0^+$) reflects the assumption about adiabatic switching the interatomic interaction on and off at $|t| \to \infty$ according to $\sim \exp(-\gamma|t|/2)$.

Following a more careful consideration of two-center ionization, including also the channel of spontaneous radiative decay of the excited state of atom B and going beyond the standard perturbation theory, we should replace the infinitesimally small parameter γ in (2.19) by the finite total decay width $\Gamma^{\Delta m}$ of the intermediate state (2.11), which accounts for the finite lifetime of this state, leading to

$$S_{2C}^{\Delta m}(\boldsymbol{b}_B) = \frac{-i\mathcal{M}_{AB}^{\Delta m}}{\omega_A - \omega_B + i\Gamma^{\Delta m}/2} \int_{-\infty}^{\infty} dt\, \mathcal{M}_B^{\Delta m}(\boldsymbol{b}_B, t) e^{i\omega_A t}. \tag{2.20}$$

The total width $\Gamma^{\Delta m}$ consists of the radiative width Γ_r^B and two-center autoionization width $\Gamma_a^{\Delta m}$ according to

$$\Gamma^{\Delta m} = \Gamma_r^B + \Gamma_a^{\Delta m}. \tag{2.21}$$

In (2.21), the radiative width due to the spontaneous radiative decay of the excited state χ_e is given by

$$\Gamma_r^B = \frac{4\omega_B^3}{3c^3} |\langle \chi_e | \boldsymbol{\xi} | \chi_g \rangle|^2 \tag{2.22}$$

and the two-center autoionization width arising due to the nonradiative decay of the excited state χ_e via interatomic electron correlations reads

$$\Gamma_a^{\Delta m} = \frac{k_r}{(2\pi)^2} \int d\Omega_k\, |\mathcal{M}_{AB}^{\Delta m}(k_r)|^2 \tag{2.23}$$

with $k_r = \sqrt{2(\omega_B + \varepsilon_g)}$ the absolute value of the momentum of the emitted electron evaluated at the resonance and Ω_k the solid angle for electron emission.

In fact, it is more reasonable to consider the amplitude (2.20) in momentum space. In general, for collisions between the projectile ion P and the $A - B$ system, the Fourier transform of the transition amplitude to momentum space and its inverse can be written as

$$\tilde{\mathcal{S}}(\boldsymbol{q}_\perp) = \frac{1}{2\pi}\int d^2\boldsymbol{b}\,\mathcal{S}(\boldsymbol{b})e^{i\boldsymbol{q}_\perp\cdot\boldsymbol{b}},$$
$$\mathcal{S}(\boldsymbol{b}) = \frac{1}{2\pi}\int d^2\boldsymbol{q}_\perp\,\tilde{\mathcal{S}}(\boldsymbol{q}_\perp)e^{-i\boldsymbol{q}_\perp\cdot\boldsymbol{b}}. \tag{2.24}$$

Here, $\boldsymbol{q}_\perp$ describes the perpendicular part of the momentum transfer between the projectile P and the $A-B$ system in the collision.

Using the first equation in (2.24), the amplitude for two-center impact ionization in momentum space is obtained to be

$$\tilde{\mathcal{S}}_{2C}^{\Delta m}(\boldsymbol{q}_\perp) = \frac{1}{2\pi}\int d^2\boldsymbol{b}_B\,\mathcal{S}_{2C}^{\Delta m}(\boldsymbol{b}_B)e^{i\boldsymbol{q}_\perp\cdot\boldsymbol{b}_B}. \tag{2.25}$$

Now, inserting (2.20) into (2.25), we arrive at

$$\tilde{\mathcal{S}}_{2C}^{\Delta m}(\boldsymbol{q}_\perp) = \frac{1}{2\pi}\frac{-i\mathcal{M}_{AB}^{\Delta m}}{\omega_A-\omega_B+i\Gamma^{\Delta m}/2}\int d^2\boldsymbol{b}_B\,e^{i\boldsymbol{q}_\perp\cdot\boldsymbol{b}_B}$$
$$\int_{-\infty}^{\infty} dt\,\mathcal{M}_B^{\Delta m}(\boldsymbol{b}_B,t)e^{i\omega_A t}. \tag{2.26}$$

In order to continue, we have to specify the matrix element $\mathcal{M}_B^{\Delta m}$, which is defined in (2.17) and enters equation (2.26). To do so, we apply the interaction $\hat{W}_B$ from (2.5), where the vector potential $\boldsymbol{A}_B$ is expressed via the scalar potential ϕ_B according to (2.6). Then, the matrix element $\mathcal{M}_B^{\Delta m}$ becomes

$$\mathcal{M}_B^{\Delta m} = \left\langle \chi_e \middle| \frac{\phi_B}{c^2}\boldsymbol{v}\cdot\hat{\boldsymbol{p}}_\xi + \frac{1}{2c^2}\boldsymbol{v}\cdot\left[\hat{\boldsymbol{p}}_\xi\phi_B\right] - \phi_B \middle| \chi_g \right\rangle. \tag{2.27}$$

We want to remind that the Liénard-Wiechert potentials from (2.6) satisfy the Lorenz condition $\partial_\mu A_B^\mu = 0$ with the four-potential $A_B^\mu = (\phi_B, \boldsymbol{A}_B)$. It is easy to show that this condition can be rewritten as $\boldsymbol{v}\cdot\left[\hat{\boldsymbol{p}}_\xi\phi_B\right] = i\frac{\partial\phi_B}{\partial t}$. Inserting the latter term into (2.27) provides

$$\mathcal{M}_B^{\Delta m} = \left\langle \chi_e \middle| \frac{\phi_B}{c^2}\boldsymbol{v}\cdot\hat{\boldsymbol{p}}_\xi + \frac{i}{2c^2}\frac{\partial\phi_B}{\partial t} - \phi_B \middle| \chi_g \right\rangle. \tag{2.28}$$

Next, we insert the matrix element (2.28) into the amplitude (2.26) and get

$$\tilde{S}_{2C}^{\Delta m}(\boldsymbol{q}_\perp) = \frac{1}{2\pi} \frac{-i\mathcal{M}_{AB}^{\Delta m}}{\omega_A - \omega_B + i\Gamma^{\Delta m}/2} \int d^2\boldsymbol{b}_B \, e^{i\boldsymbol{q}_\perp \cdot \boldsymbol{b}_B}$$
$$\times \left\langle \chi_e \middle| \int_{-\infty}^{\infty} dt \left\{ \frac{\phi_B}{c^2} \boldsymbol{v} \cdot \hat{\boldsymbol{p}}_\xi + \frac{i}{2c^2} \frac{\partial \phi_B}{\partial t} - \phi_B \right\} e^{i\omega_A t} \middle| \chi_g \right\rangle . \tag{2.29}$$

Using integration by parts for the second time integral in (2.29), where we suppose that the scalar potential ϕ_B vanishes at the boundaries $t = \pm\infty$, yields

$$\tilde{S}_{2C}^{\Delta m}(\boldsymbol{q}_\perp) = \frac{1}{2\pi} \frac{-i\mathcal{M}_{AB}^{\Delta m}}{\omega_A - \omega_B + i\Gamma^{\Delta m}/2} \int d^2\boldsymbol{b}_B \, e^{i\boldsymbol{q}_\perp \cdot \boldsymbol{b}_B}$$
$$\times \int_{-\infty}^{\infty} dt \left\langle \chi_e \middle| \phi_B \left(\frac{\boldsymbol{v} \cdot \hat{\boldsymbol{p}}_\xi}{c^2} + \frac{\omega_A}{2c^2} - 1 \right) \middle| \chi_g \right\rangle e^{i\omega_A t} . \tag{2.30}$$

In (2.30), we express the remaining matrix element in its explicit form as a space integral over the coordinate $\boldsymbol{\xi}$ and insert the scalar potential ϕ_B from (2.6). Subsequently, the amplitude $\tilde{S}_{2C}^{\Delta m}$ reads

$$\tilde{S}_{2C}^{\Delta m}(\boldsymbol{q}_\perp) = \frac{\gamma Z_P}{2\pi} \frac{-i\mathcal{M}_{AB}^{\Delta m}}{\omega_A - \omega_B + i\Gamma^{\Delta m}/2} \int d^2\boldsymbol{b}_B \int_{-\infty}^{\infty} dt \int d^3\boldsymbol{\xi} \, e^{i\boldsymbol{q}_\perp \cdot \boldsymbol{b}_B} e^{i\omega_A t}$$
$$\times \chi_e^* \frac{\frac{\boldsymbol{v} \cdot \hat{\boldsymbol{p}}_\xi}{c^2} + \frac{\omega_A}{2c^2} - 1}{\sqrt{(\boldsymbol{\xi}_\perp - \boldsymbol{b}_B)^2 + \gamma^2 (\xi_\parallel - vt)^2}} \chi_g . \tag{2.31}$$

Applying some simple manipulations, (2.31) can also be written in the form

$$\tilde{S}_{2C}^{\Delta m}(\boldsymbol{q}_\perp) = \frac{Z_P}{2\pi v} \frac{-i\mathcal{M}_{AB}^{\Delta m}}{\omega_A - \omega_B + i\Gamma^{\Delta m}/2} \int d^2\boldsymbol{b}_B \int_{-\infty}^{\infty} d(\gamma v t) \int d^3\boldsymbol{\xi}$$
$$e^{-i[\boldsymbol{q}_\perp \cdot (\boldsymbol{\xi}_\perp - \boldsymbol{b}_B) + \frac{\omega_A}{\gamma v} \gamma (\xi_\parallel - vt)]}$$
$$\chi_e^* \frac{e^{i[\boldsymbol{q}_\perp \cdot \boldsymbol{\xi}_\perp + \frac{\omega_A}{\gamma v} \gamma \xi_\parallel]} \left(\frac{\boldsymbol{v} \cdot \hat{\boldsymbol{p}}_\xi}{c^2} + \frac{\omega_A}{2c^2} - 1 \right)}{\sqrt{(\boldsymbol{\xi}_\perp - \boldsymbol{b}_B)^2 + \gamma^2 (\xi_\parallel - vt)^2}} \chi_g . \tag{2.32}$$

Taking into consideration that $\boldsymbol{s}'_B(t) = \left(\boldsymbol{\xi}_\perp - \boldsymbol{b}_B, \gamma(\xi_\parallel - vt)\right) = \left(\boldsymbol{s}'_{B\perp}, s'_{B\parallel}\right)$, we can rewrite the integrals over $\boldsymbol{b}_B$ and $\gamma v t$ in (2.32) into integrals over $\boldsymbol{s}'_{B\perp}$ and $s'_{B\parallel}$, respectively. Further, we define the vectors

$$q = \left(q_\perp, \frac{\omega_A}{v}\right) = (q_\perp, q_\parallel),$$
$$q' = \left(q_\perp, \frac{\omega_A}{\gamma v}\right) = (q_\perp, q'_\parallel), \tag{2.33}$$

where $\boldsymbol{q}$ and $\boldsymbol{q}'$ describe the momentum transfer from the projectile ion P to the target atom B as viewed in the rest frame of the target and projectile, respectively. Then, the amplitude (2.32) becomes

$$\tilde{S}_{2C}^{\Delta m}(q_\perp) = \frac{Z_P}{2\pi v} \frac{-i\mathcal{M}_{AB}^{\Delta m}}{\omega_A - \omega_B + i\Gamma^{\Delta m}/2} \int d^2 s'_{B\perp} \int_{-\infty}^{\infty} ds'_{B\parallel} \int d^3\xi \; e^{-i[q_\perp \cdot s'_{B\perp} + q'_\parallel s'_{B\parallel}]}$$
$$\times \chi_e^* \frac{e^{i[q_\perp \cdot \xi_\perp + q_\parallel \xi_\parallel]}\left(\frac{v \cdot \hat{p}_\xi}{c^2} + \frac{\omega_A}{2c^2} - 1\right)}{\sqrt{s'^2_{B\perp} + s'^2_{B\parallel}}} \chi_g, \tag{2.34}$$

which can be further simplified to two decoupled three-dimensional integrals according to

$$\tilde{S}_{2C}^{\Delta m}(q_\perp) = \frac{Z_P}{2\pi v} \frac{-i\mathcal{M}_{AB}^{\Delta m}}{\omega_A - \omega_B + i\Gamma^{\Delta m}/2}$$
$$\times \int d^3 s'_B \frac{e^{-iq' \cdot s'_B}}{s'_B}$$
$$\int d^3\xi \; \chi_e^* e^{iq \cdot \xi}\left(\frac{v \cdot \hat{p}_\xi}{c^2} + \frac{\omega_A}{2c^2} - 1\right)\chi_g. \tag{2.35}$$

In (2.35), the s'_B-integral can be calculated straightforwardly and its solution is given by

$$\int d^3 s'_B \frac{e^{-iq' \cdot s'_B}}{s'_B} = \frac{4\pi}{q'^2}. \tag{2.36}$$

Besides, we introduce the notation

$$\mathcal{F}_{eg}^{\Delta m}(q) = \left\langle \chi_e \middle| e^{iq \cdot \xi}\left(\frac{v \cdot \hat{p}_\xi}{c^2} + \frac{\omega_A}{2c^2} - 1\right) \middle| \chi_g \right\rangle$$
$$= \int d^3\xi \; \chi_e^* e^{iq \cdot \xi}\left(\frac{v \cdot \hat{p}_\xi}{c^2} + \frac{\omega_A}{2c^2} - 1\right)\chi_g. \tag{2.37}$$

Employing (2.36) and (2.37), the transition amplitude in (2.35) reads

$$\tilde{S}_{2C}^{\Delta m}(\boldsymbol{q}_\perp) = \frac{2Z_P}{i v q'^2} \frac{\mathcal{M}_{AB}^{\Delta m} \mathcal{F}_{eg}^{\Delta m}(\boldsymbol{q})}{\omega_A - \omega_B + i\Gamma^{\Delta m}/2}. \tag{2.38}$$

Finally, using (2.38) and taking into account that $\Delta m \in \{0, \pm 1\}$, the total transition amplitude for two-center impact ionization in momentum space is obtained to be

$$\tilde{S}_{2C}(\boldsymbol{q}_\perp) = \sum_{\Delta m=-1}^{1} \tilde{S}_{2C}^{\Delta m}(\boldsymbol{q}_\perp) = \frac{2Z_P}{i v q'^2} \sum_{\Delta m=-1}^{1} \frac{\mathcal{M}_{AB}^{\Delta m} \mathcal{F}_{eg}^{\Delta m}(\boldsymbol{q})}{\omega_A - \omega_B + i\Gamma^{\Delta m}/2}. \tag{2.39}$$

2.1.3 Amplitudes for Direct Impact Ionization of Atoms *A* and *B*

Within the first order of time dependent perturbation theory, the transition amplitude $\mathcal{S}_D^A(\boldsymbol{b}_A)$ for direct impact ionization of atom A in (2.12) can be written as

$$\mathcal{S}_D^A(\boldsymbol{b}_A) = \frac{1}{i} \int_{-\infty}^{\infty} dt \, \langle \phi_{\boldsymbol{k}} | \hat{W}_A(\boldsymbol{b}_A, t) | \phi_g \rangle e^{i\omega_A t} \tag{2.40}$$

with the interaction $\hat{W}_A$ between the projectile ion P and atom A given by (2.5). Again, it is more convenient to consider the amplitude (2.40) in momentum space by taking advantage of the first equation in (2.24), which provides

$$\tilde{\mathcal{S}}_D^A(\boldsymbol{q}_\perp) = \frac{1}{2\pi} \int d^2\boldsymbol{b}_A \, \mathcal{S}_D^A(\boldsymbol{b}_A) e^{i\boldsymbol{q}_\perp \cdot \boldsymbol{b}_A}. \tag{2.41}$$

At this point, one can perform a quite similar calculation to (2.25) - (2.38) in the previous Section and the resulting transition amplitude for direct impact ionization of atom A in momentum space is given by

$$\tilde{\mathcal{S}}_D^A(\boldsymbol{q}_\perp) = \frac{2Z_P}{i v q'^2} \mathcal{F}_{\boldsymbol{k}g}^{\Delta m}(\boldsymbol{q}) e^{i\boldsymbol{q}\cdot\boldsymbol{R}}, \tag{2.42}$$

where

$$\mathcal{F}_{\boldsymbol{k}g}^{\Delta m}(\boldsymbol{q}) = \left\langle \phi_{\boldsymbol{k}} \middle| e^{i\boldsymbol{q}\cdot\boldsymbol{r}} \left(\frac{\boldsymbol{v} \cdot \hat{\boldsymbol{p}}_{\boldsymbol{r}}}{c^2} + \frac{\omega_A}{2c^2} - 1 \right) \middle| \phi_g \right\rangle. \tag{2.43}$$

We note that in (2.42), $\boldsymbol{q}$ and $\boldsymbol{q}'$, which are formally defined as in (2.33), now refer to the momentum transfer from the projectile ion P to the target atom A (instead of B) as viewed in the rest frame of the target and projectile, respectively.

Accordingly, the transition amplitude $\mathcal{S}_D^B(\boldsymbol{b}_B)$ for direct impact ionization of atom B in the first order of time dependent perturbation theory is determined by

$$\mathcal{S}_D^B(\boldsymbol{b}_B) = \frac{1}{i}\int_{-\infty}^{\infty} dt\ \langle \chi_\kappa | \hat{W}_B(\boldsymbol{b}_B, t) | \chi_g \rangle e^{i\omega_B^{\text{ion}} t}. \tag{2.44}$$

Here, the interaction $\hat{W}_B$ between the projectile ion P and atom B is given in (2.5) and $\omega_B^{\text{ion}} = \epsilon_\kappa - \epsilon_g$ is the transition energy for the bound-continuum transition in B. Applying the first equation in (2.24), the amplitude (2.44) in momentum space reads

$$\tilde{\mathcal{S}}_D^B(\boldsymbol{q}_\perp) = \frac{1}{2\pi}\int d^2\boldsymbol{b}_B\ \mathcal{S}_D^B(\boldsymbol{b}_B) e^{i\boldsymbol{q}_\perp \cdot \boldsymbol{b}_B}. \tag{2.45}$$

As before, one can perform a rather similar calculation to (2.25) - (2.38) in the previous Section. Then, the transition amplitude for direct impact ionization of atom B in momentum space is obtained to be

$$\tilde{\mathcal{S}}_D^B(\boldsymbol{q}_\perp) = \frac{2Z_P}{iv q_B'^2}\mathcal{F}_{\kappa g}^{\Delta m}(\boldsymbol{q}_B) \tag{2.46}$$

with

$$\mathcal{F}_{\kappa g}^{\Delta m}(\boldsymbol{q}_B) = \left\langle \chi_\kappa \left| e^{i\boldsymbol{q}_B \cdot \boldsymbol{\xi}} \left(\frac{\boldsymbol{v} \cdot \hat{\boldsymbol{p}}_\xi}{c^2} + \frac{\omega_B^{\text{ion}}}{2c^2} - 1 \right) \right| \chi_g \right\rangle \tag{2.47}$$

and

$$\begin{aligned} \boldsymbol{q}_B &= \left(\boldsymbol{q}_\perp, \frac{\omega_B^{\text{ion}}}{v} \right), \\ \boldsymbol{q}_B' &= \left(\boldsymbol{q}_\perp, \frac{\omega_B^{\text{ion}}}{\gamma v} \right), \end{aligned} \tag{2.48}$$

where $\boldsymbol{q}_B$ and $\boldsymbol{q}_B'$ are the momenta transferred in the $P - B$ collision resulting in ionization of atom B as viewed in the rest frame of the target and projectile, respectively.

2.1.4 Cross Sections for Impact Ionization of Atom *A*

The spectrum of electrons emitted from atom A is characterized by the cross section differential in the electron momentum

$$\frac{d^3\sigma_{D+2C}}{dk^3} = \frac{1}{(2\pi)^3}\int d^2\boldsymbol{q}_\perp \, |\tilde{S}_D^A(\boldsymbol{q}_\perp) + \tilde{S}_{2C}(\boldsymbol{q}_\perp)|^2 \tag{2.49}$$

with the integration running over the plane of perpendicular momentum transfer. The cross section in (2.49) can be divided into the sum

$$\frac{d^3\sigma_{D+2C}}{dk^3} = \frac{d^3\sigma_D^A}{dk^3} + \frac{d^3\sigma_{2C}}{dk^3} + \frac{d^3\sigma_{\text{interf.}}}{dk^3}, \tag{2.50}$$

where

$$\frac{d^3\sigma_D^A}{dk^3} = \frac{1}{(2\pi)^3}\int d^2\boldsymbol{q}_\perp \, |\tilde{S}_D^A(\boldsymbol{q}_\perp)|^2 \tag{2.51}$$

and

$$\frac{d^3\sigma_{2C}}{dk^3} = \frac{1}{(2\pi)^3}\int d^2\boldsymbol{q}_\perp \, |\tilde{S}_{2C}(\boldsymbol{q}_\perp)|^2 \tag{2.52}$$

describe the partial contributions of the direct and two-center ionization mechanisms, respectively, and the term

$$\frac{d^3\sigma_{\text{interf.}}}{dk^3} = \frac{1}{(2\pi)^3}\int d^2\boldsymbol{q}_\perp \, \left(\tilde{S}_D^A\tilde{S}_{2C}^* + (\tilde{S}_D^A)^*\tilde{S}_{2C}\right) \tag{2.53}$$

arises due to the interference between the direct and two-center ionization channels.

The resonant nature of the two-center mechanism leads to the conjecture that in the small range $\omega_B + \varepsilon_g - \Gamma^{\Delta m} \lesssim \varepsilon_k \lesssim \omega_B + \varepsilon_g + \Gamma^{\Delta m}$ of electron emission energies, centered at the resonance energy $\varepsilon_{k_r} = \omega_B + \varepsilon_g$ and having a width of a few $\Gamma^{\Delta m}$, only the second term in (2.50) will be important. Indeed, we have performed numerical calculations[1] which show that, close to the resonance, the direct term (2.51) and interference term (2.53) are several orders of magnitude smaller than the

[1] In particular, we have compared the energy emission spectra associated with the cross sections given by (2.51), (2.52) and (2.53) for ionization of the Li–He dimer by proton impact

two-center term (2.52). Moreover, these calculations also show that in the range of emission energies far away from the resonance, the direct channel is the dominant ionization mechanism and only the first term in (2.50) is important. Consequently, interference between the direct and two-center channels is expected to be overall of minor importance and the interference term (2.53) in the cross section (2.50) can, in good approximation, be neglected.

Inserting (2.42) into (2.51) and (2.39) into (2.52) yields

$$\frac{d^3\sigma_D^A}{d\boldsymbol{k}^3} = \frac{Z_P^2}{2\pi^3 v^2} \int d^2\boldsymbol{q}_\perp \frac{|\mathcal{F}_{\boldsymbol{k}g}^{\Delta m}(\boldsymbol{q})|^2}{q'^4} \tag{2.54}$$

and

$$\frac{d^3\sigma_{2C}}{d\boldsymbol{k}^3} = \frac{Z_P^2}{2\pi^3 v^2} \int d^2\boldsymbol{q}_\perp \frac{1}{q'^4} \left| \sum_{\Delta m=-1}^{1} \frac{\mathcal{M}_{AB}^{\Delta m} \mathcal{F}_{eg}^{\Delta m}(\boldsymbol{q})}{\omega_A - \omega_B + i\Gamma^{\Delta m}/2} \right|^2, \tag{2.55}$$

respectively.

Now, rewriting $d\boldsymbol{k}^3 = \sqrt{2\varepsilon_k} d\varepsilon_k d\Omega_{\boldsymbol{k}}$, the direct and two-center ionization cross section differential in the emission energy and solid angle is given by

$$\frac{d^3\sigma_D^A}{d\varepsilon_k d\Omega_{\boldsymbol{k}}} = \frac{Z_P^2 \sqrt{\varepsilon_k}}{\sqrt{2}\pi^3 v^2} \int d^2\boldsymbol{q}_\perp \frac{|\mathcal{F}_{\boldsymbol{k}g}^{\Delta m}(\boldsymbol{q})|^2}{q'^4} \tag{2.56}$$

and

$$\frac{d^3\sigma_{2C}}{d\varepsilon_k d\Omega_{\boldsymbol{k}}} = \frac{Z_P^2 \sqrt{\varepsilon_k}}{\sqrt{2}\pi^3 v^2} \int d^2\boldsymbol{q}_\perp \frac{1}{q'^4} \left| \sum_{\Delta m=-1}^{1} \frac{\mathcal{M}_{AB}^{\Delta m} \mathcal{F}_{eg}^{\Delta m}(\boldsymbol{q})}{\omega_A - \omega_B + i\Gamma^{\Delta m}/2} \right|^2, \tag{2.57}$$

respectively. Here, the quantities $\mathcal{F}_{\boldsymbol{k}g}^{\Delta m}(\boldsymbol{q})$ in (2.56) as well as $\mathcal{M}_{AB}^{\Delta m}$ and $\mathcal{F}_{eg}^{\Delta m}(\boldsymbol{q})$ in (2.57) depend on the diatomic system under consideration and they are discussed in Appendix 9.2 in the Electronic Supplementary Material.

Furthermore, the energy distribution of emitted electrons is described by the cross section differential in the emission energy, which reads

---

with electron emission from the $2s$ ground state in Li and two-center ionization involving the $1s^2 \to 1s2p$ dipole excitation transition in He.

$$\frac{d\sigma_D^A}{d\varepsilon_k} = \int d\Omega_k \, \frac{d^3\sigma_D^A}{d\varepsilon_k d\Omega_k} \tag{2.58}$$

and

$$\frac{d\sigma_{2C}}{d\varepsilon_k} = \int d\Omega_k \, \frac{d^3\sigma_{2C}}{d\varepsilon_k d\Omega_k} \tag{2.59}$$

for direct and two-center ionization, respectively.

Finally, the corresponding total cross sections can be obtained according to

$$\sigma_D^A = \int_0^\infty d\varepsilon_k \, \frac{d\sigma_D^A}{d\varepsilon_k} \tag{2.60}$$

and

$$\sigma_{2C} = \int_0^\infty d\varepsilon_k \, \frac{d\sigma_{2C}}{d\varepsilon_k}. \tag{2.61}$$

2.1.5 Cross Section for Impact Ionization of Atom B

The spectrum of electrons ejected from atom B, which solely arises due to the direct impact ionization of B, is determined by the cross section differential in the electron momentum

$$\frac{d^3\sigma_D^B}{d\kappa^3} = \frac{1}{(2\pi)^3} \int d^2\boldsymbol{q}_\perp \, |\tilde{S}_D^B(\boldsymbol{q}_\perp)|^2. \tag{2.62}$$

Substitution of (2.46) into (2.62) provides

$$\frac{d^3\sigma_D^B}{d\kappa^3} = \frac{Z_P^2}{2\pi^3 v^2} \int d^2\boldsymbol{q}_\perp \, \frac{|\mathcal{F}_{\kappa g}^{\Delta m}(\boldsymbol{q}_B)|^2}{q_B'^4}. \tag{2.63}$$

Taking advantage of $d\kappa^3 = \sqrt{2\varepsilon_\kappa} d\varepsilon_\kappa d\Omega_\kappa$, the direct ionization cross section differential in the emission energy and solid angle becomes

$$\frac{d^3\sigma_D^B}{d\varepsilon_\kappa d\Omega_\kappa} = \frac{Z_P^2\sqrt{\varepsilon_\kappa}}{\sqrt{2}\pi^3 v^2} \int d^2\boldsymbol{q}_\perp \, \frac{\left|\mathcal{F}_{\kappa g}^{\Delta m}(\boldsymbol{q}_B)\right|^2}{q_B'^4}, \tag{2.64}$$

where $\mathcal{F}_{\kappa g}^{\Delta m}(\boldsymbol{q}_B)$ depends on the diatomic system under consideration and is discussed in Appendix 9.2 in the Electronic Supplementary Material. The corresponding energy differential and total cross sections can be calculated via

$$\frac{d\sigma_D^B}{d\varepsilon_\kappa} = \int d\Omega_\kappa \, \frac{d^3\sigma_D^B}{d\varepsilon_\kappa d\Omega_\kappa} \tag{2.65}$$

and

$$\sigma_D^B = \int_0^\infty d\varepsilon_\kappa \, \frac{d\sigma_D^B}{d\varepsilon_\kappa}, \tag{2.66}$$

respectively.

2.1.6 Analytical Cross Sections for Two-Center Impact Ionization

Now, we derive simple approximate formulas for the two-center impact ionization cross sections which are expressed via local atomic quantities for the individual atoms A and B accessible from the literature. In an approximate manner, we suppose that there is only one intermediate state Ψ_{ge} of the diatomic system and thus only one dipole transition in B. Then, the two-center cross section (2.57) differential in the emission energy and solid angle simplifies to

$$\frac{d^3\sigma_{2C}^{\Delta m}}{d\varepsilon_k d\Omega_k} = \frac{Z_P^2\sqrt{\varepsilon_k}}{\sqrt{2}\pi^3 v^2}\frac{|\mathcal{M}_{AB}^{\Delta m}|^2}{(\omega_A-\omega_B)^2+(\Gamma^{\Delta m})^2/4}\int d^2\boldsymbol{q}_\perp \, \frac{|\mathcal{F}_{eg}^{\Delta m}(\boldsymbol{q})|^2}{q'^4}. \tag{2.67}$$

Using $\sqrt{\varepsilon_k} = k/\sqrt{2}$ and performing some basic manipulations, (2.67) can also be written in the form

$$\frac{d^3\sigma_{2C}^{\Delta m}}{d\varepsilon_k d\Omega_k} = \frac{k}{(2\pi)^3}\frac{|\mathcal{M}_{AB}^{\Delta m}|^2}{(\omega_A-\omega_B)^2+(\Gamma^{\Delta m})^2/4} \left(\int d^2\boldsymbol{q}_\perp \left|\frac{2Z_P\mathcal{F}_{eg}^{\Delta m}(\boldsymbol{q})}{ivq'^2}\right|^2\right). \tag{2.68}$$

Here, the last term in the brackets represents the direct excitation cross section $\sigma_{exc}^{B,\Delta m}$ for atom B by ion impact and the remaining term refers to the probability that the de-excitation of B results in ionization of A. Finally, the two-center ionization cross section differential in the emission energy and solid angle is obtained to be

$$\frac{d^3\sigma_{2C}^{\Delta m}}{d\varepsilon_k d\Omega_k} = \frac{k}{(2\pi)^3} \frac{|\mathcal{M}_{AB}^{\Delta m}|^2}{(\omega_A - \omega_B)^2 + (\Gamma^{\Delta m})^2/4} \sigma_{exc}^{B,\Delta m}. \tag{2.69}$$

For determining the energy distribution of emitted electrons, we have to integrate equation (2.69) over the solid angle Ω_k for electron emission according to

$$\frac{d\sigma_{2C}^{\Delta m}}{d\varepsilon_k} = \frac{1}{2\pi} \frac{\sigma_{exc}^{B,\Delta m}}{(\omega_A - \omega_B)^2 + (\Gamma^{\Delta m})^2/4} \left(\frac{k}{(2\pi)^2} \int d\Omega_k \, |\mathcal{M}_{AB}^{\Delta m}|^2 \right). \tag{2.70}$$

We remind that due to the resonant nature of the two-center mechanism, the cross section in (2.70) noticeably contributes to the ionization of atom A only in the tiny interval $\omega_B + \varepsilon_g - \Gamma^{\Delta m} \lesssim \varepsilon_k \lesssim \omega_B + \varepsilon_g + \Gamma^{\Delta m}$ of electron emission energies, centered at the resonance energy $\varepsilon_{k_r} = \omega_B + \varepsilon_g$ and having a width of a few $\Gamma^{\Delta m}$. Within this energy range, the last term in the brackets in (2.70) is almost constant with respect to ε_k. Thus, we may evaluate this term in very good approximation at the resonance energy ε_{k_r} corresponding to the resonant value $k_r = \sqrt{2(\omega_B + \varepsilon_g)}$ of the momentum of the emitted electron, such that it takes the form of the two-center autoionization width $\Gamma_a^{\Delta m}$ given in (2.23). Taking this into account, the two-center ionization cross section differential in the emission energy results in

$$\frac{d\sigma_{2C}^{\Delta m}}{d\varepsilon_k} = \frac{1}{2\pi} \frac{\Gamma_a^{\Delta m}}{(\omega_A - \omega_B)^2 + (\Gamma^{\Delta m})^2/4} \sigma_{exc}^{B,\Delta m}. \tag{2.71}$$

In order to obtain the total cross section for two-center ionization, we have to integrate equation (2.71) over the emission energy ε_k according to

$$\sigma_{2C}^{\Delta m} = \frac{\Gamma_a^{\Delta m} \sigma_{exc}^{B,\Delta m}}{2\pi} \int_0^\infty d\varepsilon_k \frac{1}{(\omega_A - \omega_B)^2 + (\Gamma^{\Delta m})^2/4}. \tag{2.72}$$

The integral in (2.72) is solved by substituting $u = \omega_A - \omega_B = \varepsilon_k - \varepsilon_g - \omega_B$ and afterwards taking advantage of the fact that the resulting integrand only contributes to the integral in a very narrow interval $-\Gamma^{\Delta m} \lesssim u \lesssim \Gamma^{\Delta m}$ so that the lower integration boundary $-(\varepsilon_g + \omega_B)$ can be extended to $-\infty$. Then, the total two-center ionization cross section becomes

$$\sigma_{2C}^{\Delta m} = \frac{\Gamma_a^{\Delta m}}{\Gamma^{\Delta m}} \sigma_{exc}^{B,\Delta m}. \tag{2.73}$$

It is worth mentioning that the above expression has a particularly simple physical meaning, where the total cross section for two-center ionization of atom A by ion impact is the product of the impact excitation cross section of atom B times the corresponding branching ratio between the two possible pathways (nonradiative two-center autoionization and spontaneous radiative decay) of the de-excitation of B.

To conclude this Section, we will show that the quantities $|\mathcal{M}_{AB}^{\Delta m}|^2$ and $\Gamma_a^{\Delta m}$ in (2.69), (2.71) and (2.73) can be rather simply expressed via the photoionization cross section σ_{PI}^A of atom A by a photon of frequency ω_B and the radiative width Γ_r^B of the excited state of atom B. Using (2.4) and (2.16), we obtain

$$\mathcal{M}_{AB}^{\Delta m} = e^{iR\frac{\omega_B}{c}}\Bigg[\left(\boldsymbol{M}_A^{\Delta m}\cdot\boldsymbol{M}_B^{\Delta m} - \frac{3(\boldsymbol{M}_A^{\Delta m}\cdot\boldsymbol{R})(\boldsymbol{M}_B^{\Delta m}\cdot\boldsymbol{R})}{R^2}\right)\frac{1-iR\frac{\omega_B}{c}}{R^3} - \left(\boldsymbol{M}_A^{\Delta m}\cdot\boldsymbol{M}_B^{\Delta m} - \frac{(\boldsymbol{M}_A^{\Delta m}\cdot\boldsymbol{R})(\boldsymbol{M}_B^{\Delta m}\cdot\boldsymbol{R})}{R^2}\right)\frac{\left(\frac{\omega_B}{c}\right)^2}{R}\Bigg], \tag{2.74}$$

where $\boldsymbol{M}_A^{\Delta m} = \langle\phi_{\boldsymbol{k}}|\boldsymbol{r}|\phi_g\rangle$ and $\boldsymbol{M}_B^{\Delta m} = \langle\chi_g|\boldsymbol{\xi}|\chi_e\rangle$ are the (local) dipole transition matrix elements for atoms A and B, respectively. Next, we separate the bound states of A and B into radial and angular parts according to

$$\begin{aligned}\phi_{n_A l_A m_A}(\boldsymbol{r}) &= R_{n_A}^{l_A}(r) Y_{l_A}^{m_A}(\vartheta_{\boldsymbol{r}}, \varphi_{\boldsymbol{r}}),\\ \chi_{n_B l_B m_B}(\boldsymbol{\xi}) &= R_{n_B}^{l_B}(\xi) Y_{l_B}^{m_B}(\vartheta_{\boldsymbol{\xi}}, \varphi_{\boldsymbol{\xi}}).\end{aligned} \tag{2.75}$$

Here $R_{n_j}^{l_j}$ is the radial part and $Y_{l_j}^{m_j}$ the angular part (described by the spherical harmonics) of the electronic state of atom j and (n_j, l_j, m_j) is the set of principal, orbital, and magnetic quantum numbers of j ($j = A, B$). The separation into radial and angular parts is also considered for the continuum state of the emitted electron which, accordingly, can be written as

$$\phi_{\boldsymbol{k}}(\boldsymbol{r}) = \frac{2\pi}{k}\sum_{l_A=0}^{\infty} i^{l_A} e^{-i\delta_{l_A}} R_k^{l_A}(r) \sum_{m_A=-l_A}^{l_A} Y_{l_A}^{m_A}(\vartheta_{\boldsymbol{k}}, \varphi_{\boldsymbol{k}})\big[Y_{l_A}^{m_A}(\vartheta_{\boldsymbol{r}}, \varphi_{\boldsymbol{r}})\big]^*, \tag{2.76}$$

where $R_k^{l_A}$ is the radial function of the continuum state and $e^{-i\delta_{l_A}}$ is a phase factor. In (2.76), we only keep the term in l_A leading to the strongest dipole-allowed bound-continuum transition in atom A. Inserting the states (2.75) and (2.76) into (2.74) and performing the integrations over all the angles $\vartheta_{\boldsymbol{r}}, \varphi_{\boldsymbol{r}}, \vartheta_{\boldsymbol{\xi}}, \varphi_{\boldsymbol{\xi}}$, we get

$$|\mathcal{M}_{AB}^{\Delta m}|^2 = \frac{8\pi}{27}\mathcal{A}_{\Delta m}(\boldsymbol{R}, \Omega_k, \omega_B)\frac{r_A^2 r_B^2}{k^2}. \tag{2.77}$$

Furthermore, substituting (2.77) into (2.23) and calculating the Ω_k−integral, the two-center autoionization width reads

$$\Gamma_a^{\Delta m} = \frac{8\pi}{27}\mathcal{B}_{\Delta m}(\boldsymbol{R}, \omega_B)\frac{(r_A^2)_{k=k_r} r_B^2}{k_r}. \tag{2.78}$$

In (2.77) and (2.78), $r_A = \int_0^\infty dr\, r^3 R_k^{l'_A} R_{n_A}^{l_A}$ and $r_B = \int_0^\infty d\xi\, \xi^3 R_{n'_B}^{l'_B} R_{n_B}^{l_B}$ are the radial matrix elements for the ionization of A and the de-excitation of B, respectively. Besides, $\mathcal{A}_{\Delta m}$ and $\mathcal{B}_{\Delta m}$ are geometric factors, depending on the internal structure of the two-center system, which are discussed in Appendix 9.3 in the Electronic Supplementary Material. Now, we can express r_A^2 and r_B^2 via the photoionization cross section σ_{PI}^A of A by a photon of frequency ω_B and the radiative width Γ_r^B of the excited state of B, respectively, according to

$$\begin{aligned} r_A^2 &= \frac{3}{2\pi}\frac{ck}{\omega_B}\sigma_{PI}^A(\omega_B), \\ r_B^2 &= \frac{9}{4}\left(\frac{c}{\omega_B}\right)^3 \Gamma_r^B. \end{aligned} \tag{2.79}$$

Finally, inserting (2.79) into (2.77) and (2.78), we obtain

$$|\mathcal{M}_{AB}^{\Delta m}|^2 = \mathcal{A}_{\Delta m}(\boldsymbol{R}, \Omega_k, \omega_B)\frac{1}{k}\left(\frac{c}{\omega_B}\right)^4 \Gamma_r^B \sigma_{PI}^A(\omega_B) \tag{2.80}$$

and

$$\Gamma_a^{\Delta m} = \mathcal{B}_{\Delta m}(\boldsymbol{R}, \omega_B)\left(\frac{c}{\omega_B}\right)^4 \Gamma_r^B \sigma_{PI}^A(\omega_B). \tag{2.81}$$

2.2 Numerical Results and Discussion

In this Section, we present the results of numerical calculations for direct and two-center impact ionization cross sections based on the theoretical treatment considered in Section 2.1.

2.2.1 Properties of the Relativistic Charged Projectiles

In what follows, we set the projectile charge $Z_P = 1$ and thus consider the single ionization of a diatomic system by proton impact. Within the first order of perturbation theory, ionization cross sections for projectile-target collisions depend on the projectile charge as Z_P^2 and are independent of the projectile mass. For this reason, as long as the first order perturbative condition $Z_P/v \ll 1$ is satisfied, the numerical results calculated for projectiles with $Z_P = 1$ can easily be generalized to collisions involving bare ions with $Z_P > 1$.

Furthermore, concerning impact ionization by relativistic electrons, the momentum und energy transfers from the electron to the target atom are negligibly small compared with the initial momentum and energy of the projectile electron. Besides, the projectile electron and atomic electrons have essentially no overlap in the phase space. Consequently, the results obtained for collisions with proton projectiles can directly be applied also to collisions with electron (or positron) projectiles.

2.2.2 Properties of the Diatomic Targets: The Li–He and Ne–He Dimers

As diatomic targets, we may consider two heteroatomic Van-der-Waals molecules namely the Li–He dimer[2] and the Ne–He dimer.

In Li–He, both atoms are very weakly bound by the Van-der-Waals force resulting in a binding energy of just ≈ 0.5 μeV [54] which is considerably smaller than the first ionization potentials of Li ($I_A = -\varepsilon_g = 5.39$ eV) and He ($I_B = -\epsilon_g = 24.59$ eV). The mean distance between Li and He is rather large at ≈ 28 Å (≈ 53 a.u.) [54] while their equilibrium distance is ≈ 6 Å (≈ 11 a.u.) [55]. For the two-center ionization of the Li–He dimer, we only take into account the channel involving the $1s^2 \rightarrow 1s2p$ transition in He with an energy $\omega_B = 21.22$ eV, which is the first and strongest dipole-allowed transition in He. The resonance energy of the emitted electron corresponding to this transition is given by $\varepsilon_{k_r} = \omega_B + \varepsilon_g = 15.83$ eV. In addition, the radiative width of the excited $1s2p$ state in He is $\Gamma_r^B = 1.18 \times 10^{-6}$ eV [56] and the partial photoionization cross section for the $2s$ subshell in Li determined at ω_B is $\sigma_{PI}^A(21.22\text{ eV}) = 4.68 \times 10^{-19}$ cm^2 [57].

---

[2] There exist two Li–He dimers, ^{7}Li–He and ^{6}Li–He (with binding energies of 5.6 mK and 1.5 mK, respectively), the former of which is about four times stronger bound than the latter. Because ^{7}Li is much more abundant on Earth than ^{6}Li, we only consider ^{7}Li–He dimers in this thesis.

In Ne–He, both atoms are weakly bound by the Van-der-Waals force with a binding energy of ≈ 2 meV [46] that is four orders of magnitude smaller than the first ionization potentials of Ne ($I_A = -\varepsilon_g = 21.56$ eV) and He ($I_B = -\epsilon_g = 24.59$ eV). The equilibrium distance between Ne and He is ≈ 3 Å (≈ 5.7 a.u.) [46] and the mean distance is close to the equilibrium one. Concerning two-center ionization of the Ne–He dimer, we only consider the channel based on the $1s^2 \rightarrow 1s3p$ transition in He with an energy $\omega_B = 23.09$ eV. This transition is the first (and strongest) dipole-allowed transition in He for which the transition energy is larger than the ionization potential of Ne (and which has proven [46] to be highly efficient for photoionization of the Ne–He dimer). The associated resonant electron emission energy is $\varepsilon_{k_r} = \omega_B + \varepsilon_g = 1.52$ eV. Besides, the radiative width of the excited $1s3p$ state in He is given by $\Gamma_r^B = 3.73 \times 10^{-7}$ eV [56] and the photoionization cross section for Ne evaluated at ω_B is $\sigma_{PI}^A(23.09 \text{ eV}) = 8.14 \times 10^{-18}$ cm^2 [57].

In contrast to Ne–He, the mean size of Li–He (≈ 53 a.u.) significantly differs from its equilibrium size (≈ 11 a.u.). Due to the strong dependence of the two-center cross sections on the interatomic distance R, the question naturally arises which values of R should be taken in order to provide theoretical predictions allowing for experimental verification. In case of reactions involving fast electronic transitions in a dimer resulting in its breakup, the measurement of the kinetic energies of the reaction fragments often enables one to determine fairly accurate values for the magnitude of the distance R at which the process took place. Indeed, based on the results of a recent study [58], we can expect that the most likely outcome of two-center impact ionization, which we consider in this work, will be a breakup of the Li–He dimer into Li^+ and He fragments. But it can also be concluded from the results in [58] that there is no one-to-one correspondence between the kinetic energies of the fragments and the interatomic distance R.

Taking all this into account, we shall only consider cross sections for the two-center ionization of the Li–He dimer which are averaged over the size of its vibrational ground state according to

$$\sigma_{aver} = \int_0^\infty dR\, \sigma(R) |\Psi_0(R)|^2. \tag{2.82}$$

Here, $\sigma(R)$ is a cross section evaluated at an interatomic distance R and $\Psi_0(R)$ is the wave function of the molecular ground state of the Li–He dimer. Using results of [59, 60], the wave function $\Psi_0(R)$ can be approximated by

$$\Psi_0(R) = \sqrt{\frac{\alpha(\kappa - 1)}{\Gamma(\kappa)}} e^{-z/2} z^{(\kappa-1)/2} \tag{2.83}$$

with $\kappa = \sqrt{\frac{8\mu D}{\alpha^2}}$, $z = \kappa e^{-\alpha(R-R_{eq})}$, $\mu = 4.62 \times 10^3$ a.u. the reduced mass of Li–He and the fitting parameters $R_{eq} = 11.9$ a.u., $D = 5.7 \times 10^{-6}$ a.u. and $\alpha = 0.43$ a.u. Note that we consider the interatomic interaction in the dipole-dipole form (2.4) which is only valid at sufficiently large interatomic distances. Consequently, the lower boundary R_{min} of the integration over R in (2.82) should effectively not be 0 but instead fulfill the condition $R_{min} \gg 1$ a.u. However, introducing such a lower bound is no major issue due to the very rapid decrease of the probability density $|\Psi_0(R)|^2$ with decreasing the interatomic distance R in the range $R \lesssim 10$ a.u. Indeed, according to our calculations, the difference between results obtained by setting $R_{min} = 1$ a.u. and $R_{min} = 10$ a.u. does not exceed $1\,\% - 16\,\%$ (depending on the type of cross section being averaged).

2.2.3 Analytical Cross Sections for He, Li and Ne

Concerning the diatomic systems Li–He and Ne–He as well as the single atoms He, Li and Ne, we have performed two different sets of numerical calculations for obtaining ionization cross sections. The first one is based on the results of the theoretical approach shown in Secs. 2.1.1–2.1.5. The second set of numerical calculations employs approximate analytical formulas for direct and two-center ionization cross sections. In particular, the cross sections for the single ionization of atoms are calculated by using the relativistic Bethe formula. The same formula can also be applied to obtain atomic excitation cross sections, the latter of which are needed in order to calculate the approximate two-center ionization cross sections given in Section 2.1.6.

The relativistic Bethe formula for cross sections for excitation and single ionization of atoms can be written as (see, e.g. [61])

$$\sigma = \frac{8\pi Z_P^2}{v^2}\left[M^2\left\{\ln\left(\frac{\gamma v}{c}\right) - \frac{v^2}{2c^2}\right\} + C\right] \tag{2.84}$$

and is known to provide quite accurate results starting with impact energies of a few MeV/u. We note that expression (2.84) was derived within the first order of time-dependent perturbation theory in the projectile-target interaction.

In (2.84), the parameters M^2 and C depend on the internal structure of the atomic target. They can be specified for the ionization from individual subshells and for discrete excitation transitions between two subshells. We have extracted experimentally determined values for M^2 for the $2s$ and $2p$ subshell ionization of Li and Ne from [62], obtaining $M^2_{\text{Li},2s} = 0.515351$ and $M^2_{\text{Ne},2p} = 1.519$, respectively. However, in the literature we could not find any experimental or theoretical data for the parameter C for these atoms. As a reasonable alternative, we have calculated the parameter C within the scope of the relativistic binary-encounter-Q (RBEQ) model from [63] by taking advantage of the known experimental values for M^2 which yields $C_{\text{Li},2s} = 3.49$ and $C_{\text{Ne},2p} = 5.89$. Considering He, there exist accurate theoretical values for the parameters M^2 and C for discrete $1s^2 \to 1snp$ $(n = 2, 3)$ excitation transitions [64] ($M^2_{\text{He},1s\to 2p} = 0.177, C_{\text{He},1s\to 2p} = 0.82825, M^2_{\text{He},1s\to 3p} = 0.0433$ and $C_{\text{He},1s\to 3p} = 0.20338$) as well as for the ionization from the $1s^2$ ground state [65] ($M^2_{\text{He},1s} = 0.489$ and $C_{\text{He},1s} = 2.763$).

It is worth mentioning that the RBEQ model also includes an analytical expression for the cross section differential in the electron emission energy ε, which reads [63]

$$\begin{aligned} \frac{d\sigma_D}{d\varepsilon} = \frac{2\pi N_T}{c^4(\beta_t^2+\beta_b^2)b'} \Bigg\{ &\frac{Q-2}{t+1}\left(\frac{1}{w+1}+\frac{1}{t-w}\right)\frac{1+2t'}{(1+t'/2)^2} \\ &+(2-Q)\left[\frac{1}{(w+1)^2}+\frac{1}{(t-w)^2}+\frac{(b')^2}{(1+t'/2)^2}\right] \\ &+\frac{W}{N_T(w+1)}\left[\ln\left(\frac{\beta_t^2}{1-\beta_t^2}\right)-\beta_t^2-\ln(2b')\right]\Bigg\}, \end{aligned} \tag{2.85}$$

where $t = E_P/I_T$, $w = \varepsilon/I_T$, $t' = E_P/c^2$, $b' = I_T/c^2$, $\beta_t = \sqrt{1-(1+t')^{-2}}$, $\beta_b = \sqrt{1-(1+b')^{-2}}$, $Q = 4I_T M^2/N_T$ and $W = Q/(w+1)^3$ with E_P the impact energy of the projectile, I_T the ionization potential of the target atom, N_T the number of bound electrons in the atomic subshell under consideration and M^2 the parameter which was discussed above. Equation (2.85) can be used in order to calculate the energy differential cross sections for the direct impact ionization of He, Li and Ne.

2.2.4 Angular Distributions

In the following Section, we evaluate the angular distribution of electrons emitted from the diatomic system having the resonance emission energy $\varepsilon_{k_r} = \omega_B + \varepsilon_g$ at which the ionization cross section is largely dominated by two-center impact ionization of atom A. (Moreover, as we will see in Section 2.2.6, the two-center channel may strongly dominate the total electron emission in the range of emission energies centered at the resonance energy and being as broad as $\delta\varepsilon_k \sim 1$ eV.)

For exploring relativistic effects in the angular distribution, in addition to relativistic calculations, we have also performed nonrelativistic calculations in which we set $c \to \infty$.

Note that according to both, the relativistic and nonrelativistic treatments, the shape of the angular distribution is determined by a subtle interplay between the amplitudes for two-center ionization involving ion impact excitation of the different magnetic substates of the excited level of atom B. Since one cannot extract these amplitudes from cross sections, the calculation method discussed in Section 2.1.6 is not applicable here. Instead, we employ the set of numerical calculations based on the results given in Section 2.1.4.

In Fig. 2.2, we show the angular distributions of electrons emitted with the resonance energy $\varepsilon_{k_r} \approx 15.83$ eV in the process of ionization of the Li–He dimer by 1 GeV protons. These distributions are described by the cross section $\frac{d^2\sigma}{d\varepsilon_k \sin\vartheta_k d\vartheta_k}$ considered as a function of the polar emission angle ϑ_k for a given (resonance) emission energy ε_k. We can draw three main conclusions from Fig. 2.2.

First of all, the angular distributions are symmetric with respect to the polar emission angle $\vartheta_k = 90°$. This feature can be explained by the fact that the two-center process is solely driven by dipole transitions (including those transitions which result in the excitation of atom B and those leading to the subsequent energy exchange between atoms B and A). It is worth mentioning that this symmetry feature is absent in the direct ionization of A (or B) where the interference between dipole and non-dipole (mainly quadrupole) transitions results in an asymmetry between the forward ($\vartheta_k \leq 90°$) and backward ($\vartheta_k \geq 90°$) semisphere of electron emission (with more electrons being ejected into the forward semisphere).

Second, at an impact energy of 1 GeV/u, corresponding to a quite moderate value of the collisional Lorentz factor of $\gamma \approx 2.1$, the shape of the angular distribution is already strongly influenced by relativistic effects. The latter enhance the electron emission into the transverse direction and reduce the emission into the longitudinal direction with respect to the collision velocity $\boldsymbol{v}$. Note that numerical calculations for orientations of the dimer other than those considered in Fig. 2.2 show that such redistributive action of the relativistic effects is present for any orientation of the dimer (although its strength depends on the particular orientation). This feature arises from the fact that in high energy ion-atom collisions, in which the motion of atomic electrons is supposed to remain nonrelativistic, the main relativistic effect is related to the flattening of the electric field created by the projectile ion, occurring at impact velocities approaching the speed of light c and being a consequence of the Lorentz contraction of electromagnetic fields. This flattening, which was discussed in detail in Section 1.3, increases the transverse field component $\boldsymbol{E}_\perp$ ($\perp \boldsymbol{v}$) and decreases the longitudinal field component $\boldsymbol{E}_\parallel$ ($\parallel \boldsymbol{v}$), thus enhancing

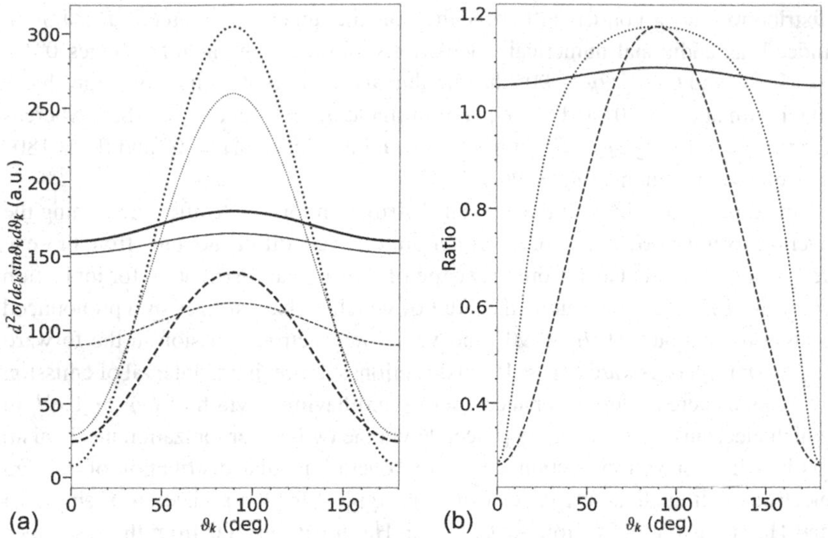

Fig. 2.2 (a) The angular distribution of electrons emitted with the resonance energy from the Li–He dimer in collisions with 1 GeV protons. The distribution was obtained by averaging over the size of the dimer and is presented for the parallel ($\boldsymbol{R} \parallel \boldsymbol{v}$, thick dashed) and perpendicular ($\boldsymbol{R} \perp \boldsymbol{v}$, thick dotted) orientation as well as for the orientational average (thick solid). In addition, the corresponding results in the nonrelativistic limit ($c \to \infty$) are depicted by thin dashed, thin dotted and thin solid curves, respectively. (b) The relativistic–to–nonrelativistic cross section ratio shown for $\boldsymbol{R} \parallel \boldsymbol{v}$ (dashed), $\boldsymbol{R} \perp \boldsymbol{v}$ (dotted) and the orientational average (solid). (This figure was originally published in Ref. [43])

electron emission in the transverse direction and reducing it in the longitudinal direction.

Third, when averaging over the orientation of the dimer, the angular distribution of emitted electrons turns out to be quite weakly dependent on the polar emission angle being almost spherically symmetric. This implies that, in case of two-center ionization, the angular momentum imparted into the initial system in the collision (via the absorption of a virtual photon) on average mainly goes to the nuclei, which results in the excitation of rotational degrees of freedom of the residual $(\text{Li–He})^+$ system.

Furthermore, it is worth mentioning that the rather pronounced maximum of the electron emission at $\vartheta_k = 90°$ for $\boldsymbol{R} \parallel \boldsymbol{v}$ and $\boldsymbol{R} \perp \boldsymbol{v}$ on the one hand as well as the very weak dependence of the electron emission on ϑ_k after averaging over the orientation of the dimer on the other hand indicate that the shape of the angular distribution has a non-trivial dependence on the angle $\theta_{\boldsymbol{R}} = \arccos(\boldsymbol{R} \cdot \boldsymbol{v}/Rv)$. Indeed, as additional numerical calculations show, for the angular ranges $0° \leq \theta_{\boldsymbol{R}} \lesssim 18°$ and $63° \lesssim \theta_{\boldsymbol{R}} \leq 90°$ the angular distribution of emitted electrons has a maximum at $\vartheta_{k} = 90°$ and two equal minima at $\vartheta_{k} = 0°$ and $\vartheta_{k} = 180°$, whereas for the range $18° \lesssim \theta_{\boldsymbol{R}} \lesssim 63°$ it has two equal maxima at $\vartheta_{k} = 0°$ and $\vartheta_{k} = 180°$ as well as a minimum at $\vartheta_{k} = 90°$.

In addition, the shape of the angular distribution for Li–He after performing the average over the orientation of the dimer qualitatively differs not only from those at $\boldsymbol{R} \parallel \boldsymbol{v}$ and $\boldsymbol{R} \perp \boldsymbol{v}$ but also from the shape of the angular distribution for ionization from an s state of a single atom, the latter of which is characterized by a pronounced emission maximum at $\vartheta_{k} \approx 90°$ and very low electron emission in the forward ($\vartheta_{k} \approx 0°$) and backward ($\vartheta_{k} \approx 180°$) directions. Hence, in the interval of emission energies centered at the resonance energy and having a width of $\delta\varepsilon_{k} \sim 1$ eV, in which electron emission mainly proceeds via the two-center ionization mechanism (as it will be shown in Section 2.2.6), the overall angular distribution of emitted electrons will qualitatively differ from that typical for the ionization of single Li and He atoms (and also from that of a Li–He dimer very far from the resonance where the two-center channel is negligible).

Note that effects similar to those discussed for the Li–He system also arise in the impact ionization of Ne–He dimers. In particular, relativistic effects related to the flattening of the electric field generated by the projectile ion tend to enhance electron emission in the transverse direction and reduce it in the longitudinal direction with respect to the collision velocity $\boldsymbol{v}$. Further, the angular distribution of electrons emitted from Ne–He via the two-center ionization channel averaged over the orientation of the dimer significantly differs from the angular emission spectra occurring in the direct impact ionization of single Ne and He atoms.

2.2.5 Energy Distributions

In this Section, we discuss the energy distribution of emitted electrons for the processes of direct and two-center impact ionization for the diatomic systems Li–He and Ne–He. In Fig. 2.3, we present the energy distribution of electrons emitted from the Li–He and Ne–He dimers via impact ionization by 1 GeV protons, which is given by the cross section $\frac{d\sigma}{d\varepsilon_k}$ evaluated as a function of the energy detuning $\delta = \omega_A - \omega_B$. More precisely, we display the two-center ionization cross section, determined by the (incoherent) sum of partial cross sections (2.71) over all intermediate states Ψ_{ge} of the diatomic system, and the cross sections for the direct ionization of Li and Ne, the latter of which were calculated by employing equation (2.85).

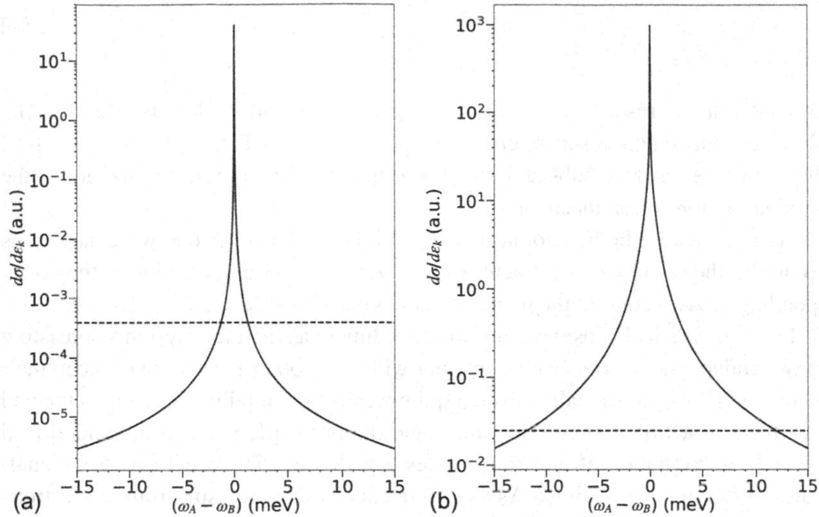

Fig. 2.3 (a) The energy distribution of emitted electrons considered as a function of the detuning $\delta = \omega_A - \omega_B$ for the Li–He dimer for two-center ionization (solid) and direct ionization of Li (dashed) in collisions with 1 GeV protons. The two-center distribution was obtained by averaging over the interatomic vector $\boldsymbol{R}$ of the dimer. (b) The corresponding energy distribution for the Ne–He dimer for two-center ionization (solid) and direct ionization of Ne (dashed). The two-center distribution was calculated by averaging over the orientation of the dimer for a fixed interatomic distance $R = 3$ Å. (This figure was originally published in Ref. [43])

It can be seen in Fig. 2.3 that the two-center cross section has a resonant structure with a maximum at the respective resonance energy $\varepsilon_{k_r} = \omega_B + \varepsilon_g$ (≈ 15.83 eV for Li–He and ≈ 1.52 eV for Ne–He). It rapidly decreases for electron emission energies $\varepsilon_k \lessgtr \varepsilon_{k_r}$, where the width of the resonance is determined by the total decay width Γ (which consists of the radiative width Γ_r^B and the two-center autoionization width Γ_a). In contrast to this, the cross sections for direct ionization of Li and Ne are only weakly dependent on the electron emission energy ε_k.

At the resonance and in a small vicinity of emission energies surrounding the resonant emission energy, the two-center ionization of atom A can exceed the direct ionization of A by several orders of magnitude. Indeed, the ratio of two-center and direct cross sections

$$\mu^{(1)} = \frac{d\sigma_{2C}/d\varepsilon_k}{d\sigma_D^A/d\varepsilon_k} \tag{2.86}$$

evaluated at the resonance yields $\approx 10^5$ for Li–He and $\approx 4 \times 10^4$ for Ne–He. However, outside the resonant energy range, $\omega_B + \varepsilon_g - \Gamma \lesssim \varepsilon_k \lesssim \omega_B + \varepsilon_g + \Gamma$, the two-center channel substantially diminishes and direct impact ionization is the dominating ionization mechanism.

We note that for both diatomic systems, Li–He and Ne–He, the two-center cross section in the relativistic treatment is about 7.5 % larger compared with the corresponding cross section in the nonrelativistic limit ($c \to \infty$).

In our theoretical consideration, we treat dimers as diatomic systems consisting of two independent atoms which interact with each other but otherwise keep their identities. However, in reality, even a quite weakly bound dimer is a molecule and thus the interaction between the dimer and the projectile ion will in general lead not only to excitation of electronic states but also of vibrational (and rotational) states of the molecular dimer. As a consequence, the energy spectrum of electrons emitted from the dimer will be split into several emission lines corresponding to the involvement of different vibrational (and rotational) states in the ionization process (see, e.g. [46, 47, 60]). Here, the electron emission lines will be rather close to each other due to the fact that the molecular states have much smaller energy separations than electronic states. For this reason, we expect that when averaging the energy spectrum of emitted electrons over the energy interval which contains all the emission lines it will correspond to that energy spectrum which is predicted by our two independent atom model of the ionization process after averaging this spectrum over the same energy interval.

2.2.6 Total Cross Sections

In the following Section, the total cross section for electron emission from the diatomic system by ion impact as a function of the projectile energy is considered. Fig. 2.4 displays the dependence of the total two-center ionization cross section, evaluated as the (incoherent) sum of partial cross sections (2.73) over all intermediate states Ψ_{ge} of the diatomic system, on the projectile energy (per nucleon) E_P for the Li–He and Ne–He dimers. Moreover, in this figure, we also show the corresponding direct ionization cross sections for single Li, Ne and He atoms.

As it can be concluded from Fig. 2.4, the two-center and direct ionization cross sections of atom A possess the same asymptotic behaviour for high projectile energies. This may be explained by two facts. First, according to equation (2.73), the

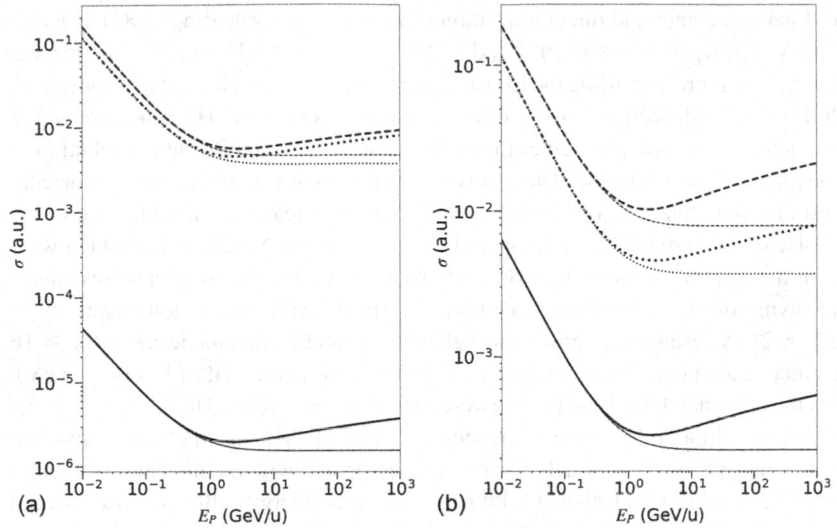

Fig. 2.4 The total cross section considered as a function of the projectile energy (per nucleon) E_p. (a) Results for two-center ionization of Li–He (thick solid) and direct ionization of Li (thick dashed) and He (thick dotted). The two-center cross section was obtained by averaging over the interatomic vector $\boldsymbol{R}$ of the dimer. In addition, the corresponding cross sections in the nonrelativistic limit ($c \to \infty$) are presented by thin solid, thin dashed and thin dotted curves, respectively. (b) Results for two-center ionization of Ne–He (thick solid) and direct ionization of Ne (thick dashed) and He (thick dotted). The two-center cross section was calculated by averaging over the orientation of the dimer at a fixed interatomic distance $R = 3$ Å. Further, the corresponding cross sections in the nonrelativistic limit are depicted by thin solid, thin dashed and thin dotted curves, respectively. Part (b) of this figure was originally published in Ref. [43]

energy dependence of the total two-center ionization cross section is solely determined by the energy dependence of the cross section for impact excitation of atom B and, second, at high impact energies the cross sections for direct impact ionization and (dipole-allowed) impact excitation have a rather similar dependence on the projectile energy.

The overall effect of two-center ionization on the total electron emission from atom A can be characterized by the ratio

$$\mu^{(2)} = \frac{\sigma_{2C}}{\sigma_D^A} \tag{2.87}$$

of total two-center and direct ionization cross sections. Evaluating (2.87) at $E_P =$ 1 GeV/u provides $\approx 3.8 \times 10^{-4}$ and $\approx 2.9 \times 10^{-2}$ for Li–He and Ne–He, respectively. Therefore, regarding the Li–He dimer, the two-center channel adds only very little to the total electron emission from Li whereas for the Ne–He dimer, which has a much smaller size, two-center ionization gives a more significant contribution to the total emission from Ne. The relative overall weakness of the two-center mechanism for the ionization of Li–He is due to two main reasons. First, the size of the Li–He dimer is much larger compared to the size of the Ne–He dimer which weakens the atom-atom interaction. Second, in case of Li–He, the two-center resonance involving the $1s^2 \rightarrow 1s2p$ dipole transition in He with a transition frequency of $\omega_B \approx 21$ eV results in the emission of electrons from Li with kinetic energy $\varepsilon_k \approx 16$ eV that is about three times larger than the ionization potential of Li ($I_A \approx 5$ eV). On the other hand, for Ne–He, the two-center resonance related to the $1s^2 \rightarrow 1s3p$ dipole transition in He with a frequency of $\omega_B \approx 23$ eV leads to electron emission with energy $\varepsilon_k \approx 2$ eV, which is very small compared with the ionization potential of Ne ($I_A \approx 22$ eV). Further, it is known that in high energy collisions with charged projectiles the majority of electrons emitted from the target have kinetic energies that do not exceed their initial atomic binding energy. Consequently, comparing the ionization of Li–He and Ne–He, the range of relatively large emission energies ~ 16 eV contributes much less to the total electron emission from Li than the range of comparatively low emission energies ~ 2 eV contributes to the total emission from Ne.

Although it has turned out that the two-center channel contributes quite less to the total electron emission from atom A, we should mention the following interesting fact. For obtaining the above discussed total cross sections, we naturally have taken into account the whole range of electron emission energies upon integration. However, if instead the integration is limited to an interval of emission energies centered at the resonance energy ε_{k_r} and having the width $\delta_{\varepsilon_k} \approx 0.5$ eV, which is much smaller than the effective width of the atomic continuum of A (~ 10 eV) but several orders of magnitude larger compared to the resonance width, then the ratio (2.87) of total two-center and direct ionization cross sections at $E_P = 1$ GeV/u becomes $\mu^{(2)} \approx 174$ and $\mu^{(2)} \approx 200$ for Li–He and Ne–He, respectively. Therefore, we can conclude that in a small ($\delta_{\varepsilon_k} \sim 1$ eV) but experimentally very well resolvable range of electron emission energies, containing the resonance energy, the two-center ionization mechanism still largely dominates the direct ionization of atom A.

Up to this point, we only have discussed electron emission from atom A (via direct and two-center impact ionization). Now, we also take into account the direct ionization of atom B by ion impact, the latter of which also contributes to the total electron emission from the whole diatomic system. The consideration of direct impact ionization of B results in a decrease of the total number of neutral atomic species B which are crucial for two-center ionization to proceed. However, this point is of minor importance as long as the condition $Z_P/v \ll 1$ is satisfied. Moreover, the direct impact ionization of B may noticeably enhance the total electron emission from the diatomic system, this way reducing the role of the two-center channel. To get an idea about the contribution of two-center ionization on the total electron emission from both atomic species A and B, we introduce the ratio

$$\mu^{(3)} = \frac{\sigma_{2C}}{\sigma_D^A + \sigma_D^B} \tag{2.88}$$

of the total two-center cross section and the sum of the direct cross sections for A and B. The ratio (2.88) is evaluated by considering only the interval of emission energies centered at the resonance energy ε_{k_r} and having the width $\delta_{\varepsilon_k} \approx 0.5$ eV when integrating over the emission energy for obtaining total cross sections. At a projectile energy of $E_P = 1$ GeV/u, we get $\mu^{(3)} \approx 8.3$ for the Li–He dimer and $\mu^{(3)} \approx 152$ for the Ne–He dimer. These values may be compared with the ratio (2.87) for the same interval of emission energies (and for the same impact energy) which is given by $\mu^{(2)} \approx 174$ for Li–He and $\mu^{(2)} \approx 200$ for Ne–He as it is already known. We can conclude from these numbers that the inclusion of electrons ejected from He to the total electron emission reduces the relative contribution of the two-center channel for both diatomic systems. For Ne–He this effect is quite weak while for Li–He the relative contribution of two-center ionization is greatly reduced (but still highly visible). To get even more information on how electron emission from He effects the total electron emission from the dimers, we look at the total cross section for direct impact ionization of He which is displayed in Fig. 2.4. Over the range of impact energies shown in this figure, the total cross section for direct ionization of Li is between 15 % and 37 % larger than that of He while the direct cross section for Ne is significantly larger compared with that for He, dominating the latter by a factor of 1.62 to 2.65. Consequently, regarding the Li–He dimer, the electron emission from He cannot be neglected whereas for the Ne–He dimer it is not too important.

The total two-center and direct impact ionization cross sections in the relativistic treatment, shown in Fig. 2.4, grow logarithmically starting with projectile energies of a few GeV/u that is a typical relativistic effect observed for any dipole-allowed transition and being once again related to the flattening of the electric field of the projectile ion, which occurs at impact velocities approaching the speed of light c. On the other hand, in the nonrelativistic limit ($c \to \infty$) the flattening of the field disappears and so does the logarithmic growth of the considered cross sections, the latter of which now simply saturate at highly relativistic impact energies.

2.2.7 Retardation Effect in Two-Center Impact Ionization

Finally, in this Section, we consider the role of the retardation effect in the diatomic system that results from the finite propagation time of the electromagnetic field which transmits the interaction between atoms A and B. In Fig. 2.5, we show the total two-center impact ionization cross section, given by the (incoherent) sum of partial cross sections (2.73) over all intermediate states Ψ_{ge} of the diatomic system, multiplied by R^6 as a function of the interatomic distance R between A and B for the Li–He dimer. When evaluating the cross section, we either use the retarded dipole-dipole interaction (2.4) or its instantaneous limit (1.6).

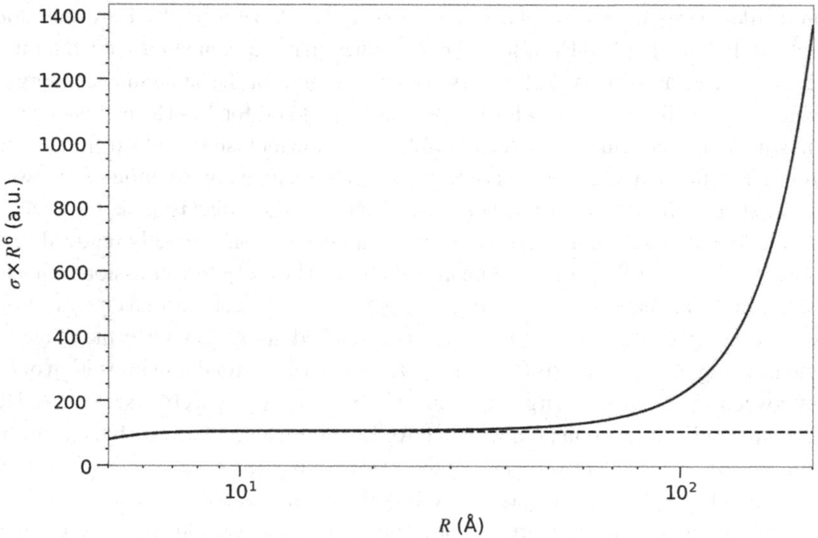

Fig. 2.5 The total cross section for the Li–He dimer multiplied by R^6 as a function of the interatomic distance R at $E_P = 1$ GeV/u using the retarded dipole-dipole interaction (2.4) (solid) and the instantaneous interaction (1.6) (dashed). The cross section is averaged over the orientation of the dimer. (This figure was originally published in Ref. [43])

Based on our detailed discussion of the retardation effect in diatomic systems in Section 1.3, this effect can be neglected as long as $T \ll \tau$, where $T = R/c$ is the time necessary for the electromagnetic field to propagate between the atoms and $\tau = 1/\omega_B$ is the electronic transition time. In this case, the field propagates essentially instantaneously and the instantaneous interaction (1.6) may be applied in very good approximation. In contrast, if $T \gg \tau$, the finite propagation of the field becomes important, the retardation effect significantly influences the interatomic interaction and the latter has to be considered in its retarded form (2.4).

Accordingly, a simple estimate for the importance of the retardation effect in diatomic systems is given by the magnitude of the ratio

$$\eta = T/\tau = \omega_B R/c \tag{2.89}$$

of the propagation time T and transition time τ. Regarding two-center ionization of the Ne–He dimer, whose size is relatively small, this ratio evaluated at $R \approx 3$ Å and $\omega_B \approx 23$ eV provides $\eta \approx 0.04$ that is quite small. Hence, the retardation effect is expected to be negligible, which also follows from our calculated total two-center cross section. Concerning two-center ionization of the Li–He dimer, whose mean size is much larger compared with that of Ne–He, the ratio (2.89) evaluated at $R \approx 28$ Å and $\omega_B \approx 21$ eV yields $\eta \approx 0.3$. For this not very small value of η one could expect a noticeable retardation effect. However, different to this simple estimate, the retardation effect on the calculated total two-center cross section in fact becomes only important at interatomic distances R which are significantly larger than the mean size of Li–He (see Fig. 2.5) and when we average over the size of the dimer, the retardation effect on the cross section gets very small being below 1 %.

Note that rather than focusing on very large dimers with relatively low transition frequencies (like the Li–He system) perhaps a more feasible way of highlighting retardation in two-center ionization of a diatomic system would be to consider ionization of a relatively small dimer but involving much larger transition frequencies. In order to get an idea of the importance of the retardation effect on the two-center ionization in such case, we suppose that the active electron in atom B undergoes a dipole-allowed transition between two atomic states, in which it is effectively restricted to the space region around the nucleus of B having a linear size a_B. Then, the dipole matrix elements of atom B will scale as a_B while its transition frequencies ω_B would scale as a_B^{-2}. Now, let $a_B \simeq 0.1$ a.u., so that $\omega_B \simeq 10^2$ a.u. In this case, according to (2.89), the retardation effect on the interatomic interaction between atom B and its neighbor atom A would become of relevance beginning already at interatomic distances R as small as $R \simeq 1$ a.u. Such simple estimate indicates that the retardation effect in two-center ionization could become important in diatomic systems where atom B contains tightly bound electrons and the two-center channel involves excitation of such an electron.

2.3 Summary and Concluding Remarks

We have considered the single electron emission from a diatomic system, consisting of two weakly bound different atomic species A and B, in relativistic collisions with charged projectiles represented by bare ions.

In systems, in which the ionization potential of atom A is smaller than an excitation energy for a dipole-allowed transition in atom B, three single ionization channels can occur: (i) direct impact ionization of A, (ii) direct impact ionization of B, and (iii) two-center impact ionization of A. Here, channels (i) and (ii) describe the well-known mechanism of direct ionization of a single atom by ion impact whereas in channel (iii) ionization of A proceeds by impact excitation of B with subsequent radiationless transfer of the excitation energy – via (long-range) interatomic electron correlations – to A, leading to its ionization.

The theoretical treatment of collisions between the $A - B$ system and the projectile was based on the semiclassical approximation, where the relative motion of the (heavy) nuclei is described classically while the active electrons are treated quantum mechanically. The semiclassical approximation is very well justified at high impact velocities. Further, the ionization channels (i)–(iii) were considered within the lowest (possible) order of the time-dependent perturbation theory. On the one hand, we obtained the transition amplitude for the direct channels by applying the first order of perturbation theory in the projectile-atom interaction. On the other hand, the transition amplitude for the two-center channel was derived by employing the second order of perturbation theory, in which both the interaction between the projectile and atom B as well as the interatomic interaction between atoms A and B are included.

We have used our theoretical approach to the ionization of diatomic systems by impact of relativistic ions in order to study single electron emission from the Li–He and Ne–He dimers. Concerning the Ne–He system, its mean size and equilibrium interatomic distance are close (both being ≈ 3 Å) and thus the calculations were carried out at a fixed value of the interatomic distance $R = 3$ Å. In contrast, regarding the Li–He system, whose mean size and equilibrium interatomic distance differ considerably, the calculations were performed by averaging cross sections obtained for a fixed interatomic distance R over the vibrational ground state of the dimer. A couple of main conclusions can be drawn from our results for the ionization of these dimers.

Relativistic effects related to the flattening of the electric field of the projectile ion in the transverse direction (with respect to the incident direction of motion of the ion), which arises as a result of the Lorentz contraction of electromagnetic fields when the collision velocity approaches the speed of light, have turned out to be significant in the ionization of a weakly bound diatomic system. However, the other type of relativistic effect of interest for us, namely the retardation in the interatomic interaction caused by the finite propagation of the electromagnetic field transmitting this interaction, has proven to be negligibly small, even for the Li–He system where one might expect a sizeable retardation effect due to the rather large mean size (≈ 28 Å) of this dimer on the atomic scale.

In two-center ionization, the relativistic effects due to the flattening of the projectile's electric field first and foremost impact the angular distribution of emitted electrons by enhancing electron emission into the transverse direction and reducing it in the longitudinal direction (counted from the collision velocity $\boldsymbol{v}$) and they become substantial already at quite small values of the Lorentz factor $\gamma \sim 1-2$. In addition, these effects increase the magnitude of the energy spectrum of emitted electrons and the total ionization cross section. However, a significant increase may only be visible at rather large values of the Lorentz factor $\gamma \gg 1$.

Besides, at the resonance energy and in its close vicinity, the two-center channel is by far the dominant ionization channel. It is so strong that it remains dominant even when considering the range of emission energies centered at the resonance energy and having a width $\delta_{\varepsilon_k} \sim 1$ eV, the latter of which is already orders of magnitude larger compared to the resonance width. Note that these findings are in accordance with the results for two-center impact ionization by nonrelativistic electrons [42].

To conclude this topic on the ionization of a weakly bound diatomic system by relativistic charged projectiles, we take a brief outlook on possible experimental verification of our theoretical predictions. The effects predicted in this study can, for instance, be tested in experiments in which the Li–He and Ne–He dimer serve as the diatomic target which is bombarded by a beam of high energy charged particles (e.g. ions) exciting especially the $1s^2 \rightarrow 1s2p$ and $1s^2 \rightarrow 1s3p$ transition in He in order to trigger efficient two-center ionization of Li and Ne, respectively. Here, it is worth mentioning that the Ne–He system (involving the $1s^2 \rightarrow 1s3p$ transition in He) was already successfully used in recent experiments on the related process of two-center resonant photoionization in a weakly bound system [46, 47] indicating that this dimer could as well be a promising candidate for experiments on two-center impact ionization.

3 Radiation- Field-Driven Ionization in Laser-Assisted Slow Atomic Collisions

This chapter provides a detailed insight into the theoretical treatment of the single electron emission in slow atomic collisions in the presence of a weak laser field via two-center resonant photoionization driven by the coupling of the colliding system to the radiation field when considering the fully relativistic interatomic interaction which accounts for the retardation effect. We derive the reaction rate for this process and compare the numerical results to those for two-center photoionization in the nonrelativistic treatment where the retardation effect, allowing for the efficient coupling to the radiation field, is not included and the interaction between the colliding atoms is regarded as instantaneous. Besides, we also discuss the relative effectiveness of two-center photoionization with respect to direct photoionization. The following chapter is mainly based on results published initially in Ref. [50].

3.1 Theoretical Consideration

3.1.1 The Coupling to the Radiation Field

First, let us consider the relativistic dispersion relation of a particle with total energy ω, momentum $\boldsymbol{q}$ and rest mass m_0, which is given by (see, e.g. [52])

$$\omega^2 - (|\boldsymbol{q}|c)^2 = (m_0 c^2)^2. \quad (3.1)$$

Supplementary Information The online version contains supplementary material available at https://doi.org/10.1007/978-3-658-43891-3_3.

A. Jacob, *Relativistic Effects in Interatomic Ionization Processes and Formation of Antimatter Ions in Interatomic Attachment Reactions*,
https://doi.org/10.1007/978-3-658-43891-3_3

The solutions of (3.1) determine the surface (the mass-shell) of a hyperboloid in energy-momentum space and are referred to as being on-mass-shell. Accordingly, if (3.1) is not satisfied, the phrase off-mass-shell is used. Equation (3.1) may also be written in terms of the particle's four-momentum $q^\mu = (\omega/c, \boldsymbol{q})$ according to

$$q^\mu q_\mu = (m_0 c)^2. \tag{3.2}$$

For a massless particle with $m_0 = 0$, as it is the case for a photon, the on-mass-shell condition (3.2) simplifies to

$$q^\mu q_\mu = 0. \tag{3.3}$$

Within the theory of Quantum Electrodynamics, atomic particles interact with each other by exchanging virtual photons whose four-momentum q^μ obeys the off-mass-shell condition

$$q^\mu q_\mu \neq 0. \tag{3.4}$$

Virtual photons are represented as inner lines between two vertices in Feynman diagrams, where they can be thought of as being emitted at one vertex and absorbed at the other. The concept of virtual photons works particularly well in systems in which the interaction is relatively weak and the interacting particles are spatially well separated. It can provide a detailed insight into the basic physics of many different processes, including Förster resonance energy transfer [30], de-excitation processes in metallic compounds [24, 25], metal oxides [66], rare gas dimers [29] and clusters [28, 67], as well as ionization reactions occurring in fast atomic collisions [68–70].

However, in some processes the kinematics of the particles allows the interaction between them to be transmitted also via the exchange of real (on-mass-shell) photons whose four-momentum q^μ satisfies condition (3.3). In particular, due to the relativistic retardation effect, taking into account the finite propagation of the interaction between the colliding particles, the coupling of the particles to the quantum radiation field becomes efficient. This enables the interaction to proceed via the exchange of on-shell photons, dramatically increasing its effective range, which may significantly affect the characteristics of the process in question. As an example, the process of electron-positron pair production in the collision of an extreme relativistic electron with an intense laser field mainly proceeds via the emission of an on-shell photon, the latter of which is converted in the laser field into an electron-positron pair [71]. In addition, the exchange of on-shell photons can strongly promote projectile-electron loss in high energy collisions with atoms [72] and excitation of ions by high energy electrons in the presence of an intense laser field [73].

It is worth mentioning that there also exists a connecting bridge between the interactions transmitted by off- and on-shell photons. For instance, the Weizsäcker-Williams approximation [74–76] exploits the fact that the electromagnetic field generated by an extreme relativistic charged particle, moving with velocities very closely approaching the speed of light, becomes almost identical to the field of an electromagnetic wave. As a consequence, the effects due to the interaction of a relativistic charged projectile with some system are in close correspondence to those effects resulting from the interaction of equivalent photons with the same system and thus the projectile's field may be replaced by equivalent photons. This approximation is well established in high energy physics (see, e.g. [52, 77, 78]).

The above examples of processes in which charged particles interact with each other by exchanging on-shell photons as well as the discussed bridging regime belong to the relativistic domain of AMO (atomic, molecular and optical) physics. However, in its low energy domain the situation looks rather different, both for processes involving weakly bound systems and collisional processes. For instance, important relaxation mechanisms that occur in weakly bound systems, like interatomic Auger [26] and Coulombic [27, 79] decay, studied in detail during the last two decades in a wide range of systems [35–38], proceed via the exchange of off-shell photons whereas the retardation effect and thus the coupling to the radiation field is unimportant for these mechanisms (see, e.g. [80]). Note that we have drawn similar conclusions in Section 2.2.7 regarding the two-step process of two-center impact ionization of a weakly bound diatomic system whose second step is represented by interatomic Coulombic decay. Here, we have seen that in the ionization of Li–He and Ne–He dimers the retardation effect and hence the coupling to the radiation field plays essentially no role. Furthermore, textbooks suggest [81–83] that the coupling to the radiation field is fully irrelevant for ionization and excitation processes taking place in (not only slow but also quite energetic) nonrelativistic atomic collisions.

It is the main goal of this study to show that—contrary to expectations—the coupling to the quantum radiation field can also strongly influence atomic processes occurring at very low energies. As an exemplary process, we will consider two-center resonant photoionization (2CPI) in slow atomic collisions, where ionization of atoms A occur in slow collisions with atoms B in the presence of a weak laser field resonantly tuned to electron transitions in B. For 2CPI in weakly bound systems, the retardation effect and therefore the coupling to the radiation field plays essentially no role [45]. However, in the following we shall see that this is not the case for 2CPI in slow collisions.

3.1.2 General Approach

Let us suppose that a beam of atomic species A (e.g. ions or atoms) moves slowly in a dilute and cold gas of atoms B, where both, A and B, are initially in their ground states. The $A - B$ system is exposed to a weak monochromatic laser field whose frequency ω is resonant to a dipole-allowed transition between the ground state with energy ϵ_g and an excited state with energy ϵ_e of B. Further, we assume that the ionization potential of atom A is smaller than the transition energy $\omega_B = \epsilon_e - \epsilon_g$ for the excitation of atom B. In such a case, laser-induced electron emission from atom A may not only occur via direct photoionization by its interaction with the laser field but also via the indirect process of two-center resonant photoionization. For a single pair of colliding atoms A and B, the latter process can proceed by the following two steps. First, B undergoes a dipole transition from the ground state with energy ϵ_g into the excited state with energy ϵ_e via (resonant) absorption of a photon from the laser field. Afterwards, B de-excites to its initial ground state and the energy excess is transferred, due to the long-range interatomic interaction, to A which, as a result, undergoes a transition from its ground state with energy ε_g into a continuum state with energy ε_k. A scheme of collisional 2CPI is shown in Fig. 3.1.

According to the standard theories (see, e.g. [81–83]) of nonrelativistic collisions of light atomic species (where all the particles involved move with velocities much smaller than the speed of light), the interaction between atoms A and B may be approximated by its instantaneous Coulomb form. In this work, however, we perform a relativistic calculation which incorporates the retardation effect accounting for the finite propagation of the electromagnetic field that transmits the interaction between A and B. This in turn allows the efficient (resonant) coupling of the $A - B$ system to the radiation field. In particular, as it follows from our consideration (and as we will see later on), the coupling to the radiation field becomes efficient in the narrow interval of resonant electron emission energies for which 2CPI proceeds via the exchange of on-shell photons whose four-momentum q^μ satisfies the on-mass-shell condition $q^\mu q_\mu = 0$.

In what follows, we use the single-electron approximation in which only one active electron in each atom A and B is considered.

Our treatment of atomic collisions is restricted to relative velocities v between A and B that are much smaller than the typical orbiting velocities $v_e \sim 1$ a.u. ($\approx 2.18 \times 10^8$ cm/s) of the active electrons in A and B. If we assume a typical

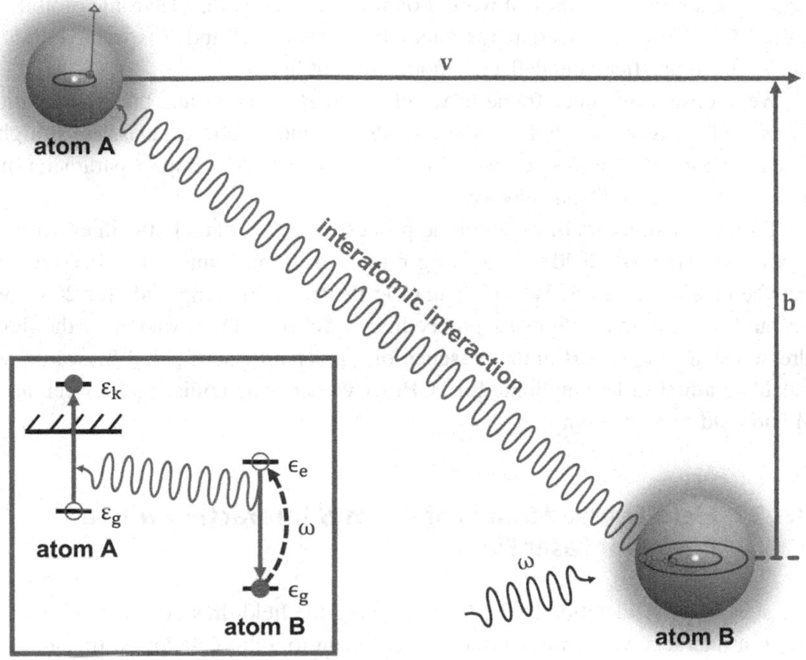

Fig. 3.1 Scheme of collisional two-center resonant photoionization (2CPI). (This figure was originally published in Ref. [50])

transition frequency $\omega_{fi} \sim 1$ a. u. for the active electrons involved in collisions of ground state atom A with ground state atom B, we obtain $\omega_{fi}/v \gg 1$. Taking into account the Massey adiabatic criterion (see, e.g. [84]), we can conclude from the above condition that the impact excitation or ionization of atom A (or B) is strongly suppressed upon collisions. Therefore, in slow collisions ($v \ll 1$ a. u.), effectively we only have to deal with direct and two-center photoionization of A.

We apply the semiclassical approximation, in which the relative motion of the heavy nuclei of A and B is treated classically while the active electrons are considered quantum mechanically. Note that this approximation is well justified starting with rather low impact energies $E \sim 1$ eV/u (see, e.g. [81]) corresponding to collision velocities $v \sim 10^{-2}$ a. u.

The overwhelming majority of electrons emitted by 2CPI via on-shell photon exchange originate from extremely distant collisions, the latter of which are of primary interest for the present work. For such collisions, the electronic orbitals of atoms A and B do not overlap, the interaction between A and B is weak and their nuclei move practically undeflected along straight lines.

We choose a reference frame in which atom B is at rest and take the position of its nucleus as the origin. In this frame, atom A moves along a classical straight-line trajectory $\boldsymbol{R}(t) = \boldsymbol{b} + \boldsymbol{v}t$, where $\boldsymbol{b} = (b_x, b_y, 0)$ is the impact parameter and $\boldsymbol{v} = (0, 0, v)$ the collision velocity.

Two interactions are involved in the process of 2CPI, namely the interaction of atom B with the laser field and the (long-range) interatomic interaction in collisions between atoms A and B. When calculating the transition amplitude for 2CPI, we account for both interactions by proceeding as follows. First, we derive the field-dressed states of atom B in the presence of a resonant laser field. Afterwards, we calculate a first-order amplitude for 2CPI by considering collisions between atom A and field-dressed atom B.

3.1.3 Field-Dressed States of Atom B Interacting with a Resonant Laser Field

We start our consideration of 2CPI by deriving the field-dressed states of atom B when it interacts with a weak laser field resonant to a dipole-allowed transition in B. The field-dressed states Ψ of B are solutions of the Schrödinger equation

$$i\frac{\partial \Psi(\boldsymbol{x}, t)}{\partial t} = (\hat{H}_B + \hat{W}_B(t))\Psi(\boldsymbol{x}, t), \tag{3.5}$$

where the coordinate $\boldsymbol{x}$ refers to the active electron in B and is given with respect to the nucleus of B and

$$\hat{H}_B = \frac{(\hat{\boldsymbol{p}}_{\boldsymbol{x}})^2}{2} - \frac{Z_B}{x} \tag{3.6}$$

is the Hamiltonian for the free (non-interacting) atom B with $\hat{\boldsymbol{p}}_{\boldsymbol{x}}$ the momentum operator for the active electron in B with respect to the nucleus of B and Z_B the effective nuclear charge of B. Further, in (3.5),

$$\hat{W}_B(t) = \frac{1}{c}\boldsymbol{A}_L(t) \cdot \hat{\boldsymbol{p}}_{\boldsymbol{x}} \tag{3.7}$$

is the interaction of atom B with the laser field, where $\boldsymbol{A}_L$ is the vector potential describing the field. We take the laser field as a classical monochromatic electromagnetic wave of linear polarization along the collision velocity $\boldsymbol{v}$ in the dipole approximation, $\boldsymbol{F}(t) = F_0 \boldsymbol{e}_z \sin(\omega t)$ with F_0 the strength of the field. In addition, we use the so-called velocity gauge, in which the electric field $\boldsymbol{F}$ is determined solely by the vector potential $\boldsymbol{A}_L$ according to $\boldsymbol{F}(t) = -\frac{1}{c}\frac{\partial \boldsymbol{A}_L(t)}{\partial t}$. Then, the (classical) vector potential $\boldsymbol{A}_L$ associated with the field $\boldsymbol{F}$ is obtained to be

$$\boldsymbol{A}_L(t) = \frac{cF_0}{\omega} \boldsymbol{e}_z \cos(\omega t). \tag{3.8}$$

We consider a dipole transition between the ground state χ_g (with energy ϵ_g) and an excited state χ_e (with energy ϵ_e) of atom B. The frequency ω of the laser field shall be resonantly tuned to the corresponding excitation energy $\omega_B = \epsilon_e - \epsilon_g$. In this case, the field-dressed (bound) states $\Psi^\pm$ of B can be written as

$$\Psi^\pm(\boldsymbol{x}, t) = a_g^\pm(t)\chi_g(\boldsymbol{x})e^{-i\epsilon_g t} + a_e^\pm(t)\chi_e(\boldsymbol{x})e^{-i\epsilon_e t}. \tag{3.9}$$

Here, $a_g^\pm(t)$ and $a_e^\pm(t)$ are time-dependent coefficients to be determined. In addition, we suppose that the laser field is switched on adiabatically at $t \to -\infty$ and set the boundary conditions $\Psi^+(\boldsymbol{x}, t \to -\infty) = \chi_g(\boldsymbol{x})e^{-i\epsilon_g t}$ (or $a_g^+(t \to -\infty) = 1$, $a_e^+(t \to -\infty) = 0$) and $\Psi^-(\boldsymbol{x}, t \to -\infty) = \chi_e(\boldsymbol{x})e^{-i\epsilon_e t}$ (or $a_g^-(t \to -\infty) = 0$, $a_e^-(t \to -\infty) = 1$).

Next, we insert (3.9) into (3.5). The resulting equation is projected on $\langle\chi_g|$ and $\langle\chi_e|$, respectively. Afterwards, we take advantage of the rotating wave approximation (see, e.g. [85]), in which the rapidly oscillating time-dependent terms are dropped. Taking all this into account, the set of equations for the unknown coefficients $a_g^\pm(t)$ and $a_e^\pm(t)$ in (3.9) reads

$$\begin{aligned} i\dot{a}_g^\pm(t) &= \mathcal{W}_{ge} a_e^\pm(t) e^{-i\Delta t}, \\ i\dot{a}_e^\pm(t) &= \mathcal{W}_{ge}^* a_g^\pm(t) e^{i\Delta t}, \end{aligned} \tag{3.10}$$

where $\Delta = (\epsilon_e - \epsilon_g) - \omega = \omega_B - \omega$ is the detuning between the excitation energy of atom B and the laser frequency,

$$\mathcal{W}_{ge} = \frac{F_0}{2\omega} \langle \chi_g | \boldsymbol{e}_z \cdot \hat{\boldsymbol{p}}_{\boldsymbol{x}} | \chi_e \rangle \tag{3.11}$$

and $\mathcal{W}_{ge}^*$ is the complex conjugate of $\mathcal{W}_{ge}$. Now, we define $\tilde{a}_e^{\pm}(t) = a_e^{\pm}(t)e^{-i\Delta t}$, where $\tilde{a}_e^{+}(t \to -\infty) = 0$ and $\tilde{a}_e^{-}(t \to -\infty) = e^{-i\Delta t}$. Then, (3.10) becomes

$$\begin{aligned} i\dot{a}_g^{\pm}(t) &= \mathcal{W}_{ge}\tilde{a}_e^{\pm}(t), \\ i\dot{\tilde{a}}_e^{\pm}(t) - \Delta\tilde{a}_e^{\pm}(t) &= \mathcal{W}_{ge}^* a_g^{\pm}(t). \end{aligned} \tag{3.12}$$

The set of equations (3.12) can be straightforwardly solved by considering stationary solutions $a_g(t) = A_g e^{-iEt}$ and $\tilde{a}_e(t) = A_e e^{-iEt}$ with $|A_g|^2 + |A_e|^2 = 1$ and using the boundary conditions for $a_g^{\pm}(t)$ and $\tilde{a}_e^{\pm}(t)$ from above. Finally, noting that $a_e^{\pm}(t) = \tilde{a}_e^{\pm}(t)e^{i\Delta t}$, the solutions for the coefficients $a_g^{\pm}(t)$ and $a_e^{\pm}(t)$, which determine the field-dressed states of atom B in (3.9), are given by

$$\begin{aligned} a_g^{+}(t) &= \frac{\mathcal{W}_{ge}}{\sqrt{E_+^2 + |\mathcal{W}_{ge}|^2}} e^{-iE_+ t} e^{-i\varphi_0}, \\ a_e^{+}(t) &= \sqrt{\frac{E_+^2}{E_+^2 + |\mathcal{W}_{ge}|^2}} e^{-i(E_+ - \Delta)t} e^{-i\varphi_0} \end{aligned} \tag{3.13}$$

and

$$\begin{aligned} a_g^{-}(t) &= \frac{\mathcal{W}_{ge}}{\sqrt{E_-^2 + |\mathcal{W}_{ge}|^2}} e^{-iE_- t}, \\ a_e^{-}(t) &= \sqrt{\frac{E_-^2}{E_-^2 + |\mathcal{W}_{ge}|^2}} e^{-i(E_- - \Delta)t} \end{aligned} \tag{3.14}$$

with $\varphi_0 = \arg \mathcal{W}_{ge}$ and $E_\pm = \frac{1}{2}\left(\Delta \mp \frac{\Delta}{|\Delta|}\sqrt{\Delta^2 + 4|\mathcal{W}_{ge}|^2}\right)$.

3.1.4 Amplitude for Two-Center Photoionization Via Coupling to the Radiation Field

In the last Section, we have obtained the states $\Psi^{\pm}$ of atom B dressed by the laser field, this way taking into account the interaction between B and the field. Now, we turn to the ionization of atom A by its collisional (long-range) interaction with field-dressed atom B.

In general, there are different ways to arrive at the transition amplitude for 2CPI. In our approach to this process, which accounts for the coupling of the $A-B$ system to the quantum radiation field, we start by considering the coupling $j^A_\mu A^\mu_B$ between the transition four-current $j^A_\mu = (c\rho_A, \boldsymbol{j_A})$ of the active electron in atom A and the four-potential $A^\mu_B = (\phi, \boldsymbol{A})$ of the field created by the other active electron in atom B. The corresponding first-order transition amplitude for the interaction between A and B can be written as (see, e.g. [86])

$$a_{2C} = -\frac{i}{c^2}\int d^4x\, j^A_\mu(x)A^\mu_B(x), \tag{3.15}$$

where $x^\mu = (ct, \boldsymbol{x})$ is the four-space-time vector.

In (3.15), we insert the inverse Fourier transforms

$$\begin{aligned} j^A_\mu(x) &= \frac{1}{(2\pi)^2}\int d^4k_A\, \tilde{j}^A_\mu(k_A)e^{-ik_Ax}, \\ A^\mu_B(x) &= \frac{1}{(2\pi)^2}\int d^4k_B\, \tilde{A}^\mu_B(k_B)e^{-ik_Bx} \end{aligned} \tag{3.16}$$

with $k^\mu_A = (\tilde{\omega}_A/c, \boldsymbol{k_A})$ and $k^\mu_B = (\tilde{\omega}_B/c, \boldsymbol{k_B})$ the four-wave vectors of the active electrons in A and B, respectively. Subsequent integration over the space-time provides

$$a_{2C} = -\frac{i}{c^2}\int d^4k_A \int d^4k_B\, \tilde{j}^A_\mu(k_A)\tilde{A}^\mu_B(k_B)\delta(k_A + k_B). \tag{3.17}$$

Next, we integrate over k_B by taking advantage of the delta function which yields

$$a_{2C} = -\frac{i}{c^2}\int d^4k_A\, \tilde{j}^A_\mu(k_A)\tilde{A}^\mu_B(-k_A). \tag{3.18}$$

The four-potential $A^\mu_B(x)$ satisfies the Maxwell equations

$$\left(\frac{1}{c^2}\frac{\partial^2}{\partial t^2} - \Delta_x\right)A^\mu_B(x) = \frac{4\pi}{c}j^\mu_B(x), \tag{3.19}$$

the latter of which can be solved in the four-dimensional k_B space leading to

$$\tilde{A}^\mu_B(k_B) = -\frac{4\pi}{c}\tilde{G}_F(k_B)\tilde{j}^\mu_B(k_B). \tag{3.20}$$

Here, $\tilde{G}_F(k_B) = \left((\tilde{\omega}_B/c)^2 - \boldsymbol{k}_B^2 + i\eta\right)^{-1}$ $(\eta \to 0^+)$ is the Feynman propagator for a massless Klein-Gordon particle. Using (3.20), the transition amplitude in (3.18) becomes

$$a_{2C} = \frac{4\pi i}{c^3} \int d^4k_A\, \tilde{G}_F(-k_A)\tilde{j}_\mu^A(k_A)\tilde{j}_B^\mu(-k_A). \tag{3.21}$$

In (3.21), the Fourier transforms of the transition four-currents are given by

$$\begin{aligned}\tilde{j}_\mu^A(k_A) &= \frac{1}{(2\pi)^2} \int d^4x\, j_\mu^A(x)e^{ik_Ax},\\ \tilde{j}_B^\mu(-k_A) &= \frac{1}{(2\pi)^2} \int d^4x\, j_B^\mu(x)e^{i(-k_A)x}.\end{aligned} \tag{3.22}$$

According to the discussion in Section 1.3, we consider a nonrelativistic electron motion and thus the four-current for the electron in atom A is determined by

$$j_\mu^A(x) = (c\rho_A, -\boldsymbol{j}_A), \tag{3.23}$$

where

$$\begin{aligned}\rho_A(x) &= \int d^3\boldsymbol{v}\, \Phi_{\boldsymbol{k}_e}^*(\boldsymbol{v},t)\left[Z_{\text{eff}}^A\delta(\boldsymbol{x}-\boldsymbol{R}(t)) - \delta(\boldsymbol{x}-\boldsymbol{v})\right]\Phi_g(\boldsymbol{v},t),\\ \boldsymbol{j}_A(x) &= \int d^3\boldsymbol{v}\, \delta(\boldsymbol{x}-\boldsymbol{v})\frac{(-1)}{2}\left\{\Phi_{\boldsymbol{k}_e}^*(\boldsymbol{v},t)\hat{\boldsymbol{p}}_{\boldsymbol{v}}\Phi_g(\boldsymbol{v},t) + \Phi_g(\boldsymbol{v},t)\hat{\boldsymbol{p}}_{\boldsymbol{v}}^*\Phi_{\boldsymbol{k}_e}^*(\boldsymbol{v},t)\right\}\end{aligned} \tag{3.24}$$

with Φ_g and $\Phi_{\boldsymbol{k}_e}$ the initial and final states of A, respectively, and Z_{eff}^A the effective charge of A. In (3.24), the $\boldsymbol{v}$-integral in the first term of $\rho_A(x)$ vanishes since the states Φ_g and $\Phi_{\boldsymbol{k}_e}$ are orthogonal. The remaining integrals over $\boldsymbol{v}$ are solved by taking advantage of the delta function $\delta(\boldsymbol{x}-\boldsymbol{v})$. Then, we arrive at

$$\begin{aligned}\rho_A(x) &= -\Phi_{\boldsymbol{k}_e}^*(\boldsymbol{x},t)\Phi_g(\boldsymbol{x},t),\\ \boldsymbol{j}_A(x) &= -\frac{1}{2}\left\{\Phi_{\boldsymbol{k}_e}^*(\boldsymbol{x},t)\hat{\boldsymbol{p}}_{\boldsymbol{x}}\Phi_g(\boldsymbol{x},t) + \Phi_g(\boldsymbol{x},t)\hat{\boldsymbol{p}}_{\boldsymbol{x}}^*\Phi_{\boldsymbol{k}_e}^*(\boldsymbol{x},t)\right\}.\end{aligned} \tag{3.25}$$

Here, the initial and final states of atom A can be written as

$$\begin{aligned}\Phi_g(\boldsymbol{x},t) &= \varphi_g(\boldsymbol{x}-\boldsymbol{R}(t))e^{-i\varepsilon_g t}\alpha(\boldsymbol{x},t),\\ \Phi_{\boldsymbol{k}_e}(\boldsymbol{x},t) &= \varphi_{\boldsymbol{k}_e}(\boldsymbol{x}-\boldsymbol{R}(t))e^{-i\varepsilon_{k_e} t}\alpha(\boldsymbol{x},t),\end{aligned} \tag{3.26}$$

where φ_g is the ground state (with energy ε_g) and $\varphi_{\boldsymbol{k}_e}$ the continuum state (with asymptotic momentum $\boldsymbol{k}_e$ and energy $\varepsilon_{k_e} = \boldsymbol{k}_e^2/2$) of A in its restframe. Further, $\alpha(\boldsymbol{x},t) = e^{i[\boldsymbol{v}\cdot(\boldsymbol{x}-\boldsymbol{R}(t))-v^2/2]}$ is the so-called translational factor that accounts for the motion of atom A in the restframe of atom B.
The corresponding four-current $j_B^{\mu}(x)$ for the electron in atom B reads

$$j_B^{\mu}(x) = (c\rho_B, \boldsymbol{j}_B) \tag{3.27}$$

with

$$\begin{aligned}\rho_B(x) &= \int d^3\boldsymbol{\nu}\; \Psi_f^*(\boldsymbol{\nu},t)\big[Z_{\text{eff}}^B\delta(\boldsymbol{x}) - \delta(\boldsymbol{x}-\boldsymbol{\nu})\big]\Psi_i(\boldsymbol{\nu},t),\\ \boldsymbol{j}_B(x) &= \int d^3\boldsymbol{\nu}\; \delta(\boldsymbol{x}-\boldsymbol{\nu})\frac{(-1)}{2}\Big\{\Psi_f^*(\boldsymbol{\nu},t)\hat{\boldsymbol{p}}_{\boldsymbol{\nu}}\Psi_i(\boldsymbol{\nu},t) + \Psi_i(\boldsymbol{\nu},t)\hat{\boldsymbol{p}}_{\boldsymbol{\nu}}^*\Psi_f^*(\boldsymbol{\nu},t)\Big\},\end{aligned} \tag{3.28}$$

where Ψ_i and Ψ_f are the initial and final states of B, respectively, and Z_{eff}^B is the effective charge of B. Taking into account that atom B is initially in its ground state when the laser field is switched on, Ψ_i and Ψ_f are determined by the field-dressed states $\Psi^{\pm}$ of B from (3.9), (3.13) and (3.14) according to $\Psi_i = \Psi^+$ and $\Psi_f = \Psi^{\pm}$. In (3.28), the $\boldsymbol{\nu}$-integral in the first term of $\rho_B(x)$ is solved by accounting for the orthonormality of the ground (χ_g) and excited (χ_e) states of B. The remaining integrals over $\boldsymbol{\nu}$ are performed by using the delta function $\delta(\boldsymbol{x}-\boldsymbol{\nu})$. Subsequently, we get

$$\begin{aligned}\rho_B(x) &= Z_{\text{eff}}^B\delta(\boldsymbol{x})\big[\big(a_g^{\pm}(t)\big)^*a_g^+(t) + \big(a_e^{\pm}(t)\big)^*a_e^+(t)\big] - (\Psi^{\pm}(\boldsymbol{x},t))^*\Psi^+(\boldsymbol{x},t),\\ \boldsymbol{j}_B(x) &= -\frac{1}{2}\Big\{\big(\Psi^{\pm}(\boldsymbol{x},t)\big)^*\hat{\boldsymbol{p}}_{\boldsymbol{x}}\Psi^+(\boldsymbol{x},t) + \Psi^+(\boldsymbol{x},t)\hat{\boldsymbol{p}}_{\boldsymbol{x}}^*\big(\Psi^{\pm}(\boldsymbol{x},t)\big)^*\Big\}.\end{aligned} \tag{3.29}$$

Next, we insert the four-current $j_{\mu}^A(x)$ of the electron in atom A, given by equations (3.23) and (3.25), into the first expression of (3.22), rewrite the integral over the space coordinate $\boldsymbol{x}$ into an integral over the coordinate $\boldsymbol{r} = \boldsymbol{x} - \boldsymbol{R}(t)$ of the electron in A with respect to the nucleus of A and afterwards perform the integration over the time t, which yields

$$\tilde{j}^A_\mu(k_A) = \Big(-\frac{c^2}{2\pi} e^{-i\boldsymbol{k}_A\cdot\boldsymbol{b}} \delta(\omega_A + \tilde{\omega}_A - \boldsymbol{k}_A \cdot \boldsymbol{v}) F^{A,0}_{\boldsymbol{k}_e g}(\boldsymbol{k}_A), \frac{c}{4\pi} e^{-i\boldsymbol{k}_A\cdot\boldsymbol{b}} \delta$$

$$(\omega_A + \tilde{\omega}_A - \boldsymbol{k}_A \cdot \boldsymbol{v}) \big\{ \boldsymbol{F}^{A,1}_{\boldsymbol{k}_e g}(\boldsymbol{k}_A) + 2\boldsymbol{v} F^{A,0}_{\boldsymbol{k}_e g}(\boldsymbol{k}_A) \big\} \Big) \tag{3.30}$$

with

$$F^{A,0}_{\boldsymbol{k}_e g}(\boldsymbol{k}_A) = \langle \varphi_{\boldsymbol{k}_e}(\boldsymbol{r}) | e^{-i\boldsymbol{k}_A\cdot\boldsymbol{r}} | \varphi_g(\boldsymbol{r}) \rangle,$$
$$\boldsymbol{F}^{A,1}_{\boldsymbol{k}_e g}(\boldsymbol{k}_A) = \langle \varphi_{\boldsymbol{k}_e}(\boldsymbol{r}) | e^{-i\boldsymbol{k}_A\cdot\boldsymbol{r}} \hat{\boldsymbol{p}}_{\boldsymbol{r}} + \hat{\boldsymbol{p}}_{\boldsymbol{r}} e^{-i\boldsymbol{k}_A\cdot\boldsymbol{r}} | \varphi_g(\boldsymbol{r}) \rangle. \tag{3.31}$$

Further, we insert the four-current $j^\mu_B(x)$ of the electron in atom B, given by equations (3.27) and (3.29), into the second expression of (3.22) and calculate in $\tilde{j}^0_B(-k_A)$ the $\boldsymbol{x}$-integral over the term which contains $\delta(\boldsymbol{x})$ by using the latter. Subsequently, we perform the integration over the time t, where we exploit the fact that $|E_+ - E_-|$ is much smaller than the transition energies in A and B. Then, we arrive at

$$(\tilde{j}^\mu_B)^\pm(-k_A) = \Bigg(\frac{c^2}{2\pi} \Big[\big\{ D^\pm_{gg} F^{B,0}_{gg}(\boldsymbol{k}_A) + D^\pm_{ee} F^{B,0}_{ee}(\boldsymbol{k}_A) \big\} \delta(\tilde{\omega}_A)$$
$$+ D^\pm_{ge} F^{B,0}_{ge}(\boldsymbol{k}_A) \delta(\tilde{\omega}_A + \omega) + D^\pm_{eg} F^{B,0}_{eg}(\boldsymbol{k}_A) \delta(\tilde{\omega}_A - \omega) \Big],$$
$$- \frac{c}{4\pi} \Big[\big\{ D^\pm_{gg} \boldsymbol{F}^{B,1}_{gg}(\boldsymbol{k}_A) + D^\pm_{ee} \boldsymbol{F}^{B,1}_{ee}(\boldsymbol{k}_A) \big\} \delta(\tilde{\omega}_A)$$
$$+ D^\pm_{ge} \boldsymbol{F}^{B,1}_{ge}(\boldsymbol{k}_A) \delta(\tilde{\omega}_A + \omega) + D^\pm_{eg} \boldsymbol{F}^{B,1}_{eg}(\boldsymbol{k}_A) \delta(\tilde{\omega}_A - \omega) \Big] \Bigg). \tag{3.32}$$

Here,

$$F^{B,0}_{\alpha\beta}(\boldsymbol{k}_A) = Z^B_{\text{eff}} \delta_{\alpha\beta} - \langle \chi_\alpha(\boldsymbol{x}) | e^{i\boldsymbol{k}_A\cdot\boldsymbol{x}} | \chi_\beta(\boldsymbol{x}) \rangle,$$
$$\boldsymbol{F}^{B,1}_{\alpha\beta}(\boldsymbol{k}_A) = \langle \chi_\alpha(\boldsymbol{x}) | e^{i\boldsymbol{k}_A\cdot\boldsymbol{x}} \hat{\boldsymbol{p}}_{\boldsymbol{x}} + \hat{\boldsymbol{p}}_{\boldsymbol{x}} e^{i\boldsymbol{k}_A\cdot\boldsymbol{x}} | \chi_\beta(\boldsymbol{x}) \rangle \tag{3.33}$$

for $\alpha\beta \in \{gg, ee, ge, eg\}$ and

$$D^{\pm}_{gg} = e^{-i\varphi_{\pm}} \frac{|\mathcal{W}_{ge}|^2}{\sqrt{(E^2_{\pm} + |\mathcal{W}_{ge}|^2)(E^2_{\mp} + |\mathcal{W}_{ge}|^2)}},$$

$$D^{\pm}_{ee} = e^{-i\varphi_{\pm}} \sqrt{\frac{E^2_{\pm} E^2_{\mp}}{(E^2_{\pm} + |\mathcal{W}_{ge}|^2)(E^2_{\mp} + |\mathcal{W}_{ge}|^2)}},$$

$$D^{\pm}_{ge} = e^{-i\varphi_{\pm}} \sqrt{\frac{E^2_{\mp}}{(E^2_{\pm} + |\mathcal{W}_{ge}|^2)(E^2_{\mp} + |\mathcal{W}_{ge}|^2)}} \mathcal{W}^*_{ge},$$

$$D^{\pm}_{eg} = e^{-i\varphi_{\pm}} \sqrt{\frac{E^2_{\pm}}{(E^2_{\pm} + |\mathcal{W}_{ge}|^2)(E^2_{\mp} + |\mathcal{W}_{ge}|^2)}} \mathcal{W}_{ge}, \tag{3.34}$$

where $\varphi_+ = 0$ and $\varphi_- = \varphi_0$.
Now, insertion of $\tilde{j}^A_\mu(k_A)$ and $(\tilde{j}^\mu_B)^\pm(-k_A)$ from (3.30) and (3.32), respectively, into (3.21) provides the amplitude $a^{\pm}_{2C}$ for the collisional interaction between atom A and field-dressed atom B. It can be separated into contributions $a^{\pm}_{gg}$ and $a^{\pm}_{ee}$ related to elastic transitions $\chi_g \to \chi_g$ and $\chi_e \to \chi_e$ in B as well as $a^{\pm}_{ge}$ and $a^{\pm}_{eg}$ referring to de-excitation and excitation transitions $\chi_e \to \chi_g$ and $\chi_g \to \chi_e$ in B:

$$a^{\pm}_{2C} = a^{\pm}_{gg} + a^{\pm}_{ee} + a^{\pm}_{ge} + a^{\pm}_{eg}. \tag{3.35}$$

Here,

$$a^{\pm}_{gg} = -\frac{ic}{\pi} D^{\pm}_{gg} \left(I^{gg,0}_1 + \frac{I^{gg,0}_2}{4c^2} + \frac{\boldsymbol{v} \cdot \boldsymbol{I}^{gg,0}_3}{2c^2} \right),$$

$$a^{\pm}_{ee} = -\frac{ic}{\pi} D^{\pm}_{ee} \left(I^{ee,0}_1 + \frac{I^{ee,0}_2}{4c^2} + \frac{\boldsymbol{v} \cdot \boldsymbol{I}^{ee,0}_3}{2c^2} \right),$$

$$a^{\pm}_{ge} = -\frac{ic}{\pi} D^{\pm}_{ge} \left(I^{ge,\omega}_1 + \frac{I^{ge,\omega}_2}{4c^2} + \frac{\boldsymbol{v} \cdot \boldsymbol{I}^{ge,\omega}_3}{2c^2} \right),$$

$$a^{\pm}_{eg} = -\frac{ic}{\pi} D^{\pm}_{eg} \left(I^{eg,-\omega}_1 + \frac{I^{eg,-\omega}_2}{4c^2} + \frac{\boldsymbol{v} \cdot \boldsymbol{I}^{eg,-\omega}_3}{2c^2} \right) \tag{3.36}$$

with

$$I_1^{\alpha\beta,\Omega} = \int d^4k_A\,\delta(\tilde{\omega}_A+\Omega)\delta(\omega_A+\tilde{\omega}_A-\boldsymbol{k}_A\cdot\boldsymbol{v})\frac{e^{-i\boldsymbol{k}_A\cdot\boldsymbol{b}}F_{\boldsymbol{k}_e g}^{A,0}(\boldsymbol{k}_A)F_{\alpha\beta}^{B,0}(\boldsymbol{k}_A)}{\left(\frac{\tilde{\omega}_A}{c}\right)^2-k_A^2+i\eta},$$

$$I_2^{\alpha\beta,\Omega} = \int d^4k_A\,\delta(\tilde{\omega}_A+\Omega)\delta(\omega_A+\tilde{\omega}_A-\boldsymbol{k}_A\cdot\boldsymbol{v})\frac{e^{-i\boldsymbol{k}_A\cdot\boldsymbol{b}}\boldsymbol{F}_{\boldsymbol{k}_e g}^{A,1}(\boldsymbol{k}_A)\cdot\boldsymbol{F}_{\alpha\beta}^{B,1}(\boldsymbol{k}_A)}{\left(\frac{\tilde{\omega}_A}{c}\right)^2-k_A^2+i\eta},$$

$$\boldsymbol{I}_3^{\alpha\beta,\Omega} = \int d^4k_A\,\delta(\tilde{\omega}_A+\Omega)\delta(\omega_A+\tilde{\omega}_A-\boldsymbol{k}_A\cdot\boldsymbol{v})\frac{e^{-i\boldsymbol{k}_A\cdot\boldsymbol{b}}F_{\boldsymbol{k}_e g}^{A,0}(\boldsymbol{k}_A)\boldsymbol{F}_{\alpha\beta}^{B,1}(\boldsymbol{k}_A)}{\left(\frac{\tilde{\omega}_A}{c}\right)^2-k_A^2+i\eta} \tag{3.37}$$

for $(\alpha\beta,\Omega) \in \{(gg,0),(ee,0),(ge,\omega),(eg,-\omega)\}$.

In the amplitude (3.35), the contribution $a_{ge}^{\pm}$ corresponds to the case, where in collisions between A and B the energy necessary for the ionization of A is gained from the energy excess upon de-excitation in B. Consequently, $a_{ge}^{\pm}$ is associated with the ionization mechanism underlying the process of 2CPI. For the other contributions in (3.35), single electron emission from A is accompanied by excitation or elastic transitions in B, such that the energy needed for ionizing A has to be provided by the relative motion of colliding atoms A and B. Since we are only interested in ionization of A via the 2CPI channel, in (3.35), we may only keep the term $a_{ge}^{\pm}$ and omit the other contributions. (Note that for collision velocities $v \ll 1$ a.u., the energy gained from the relative motion would not be sufficient to ionize A anyway.)

Now that we solely deal with de-excitation transitions in B, we introduce the simplifying notations

$$D^{\pm} = D_{ge}^{\pm} = e^{-i\varphi_{\pm}}\sqrt{\frac{E_{\mp}^2}{(E_{\pm}^2+|\mathcal{W}_{ge}|^2)(E_{\mp}^2+|\mathcal{W}_{ge}|^2)}}\mathcal{W}_{ge}^* \tag{3.38}$$

and

$$I_1 = I_1^{ge,\omega} = \int d^4k_A\,\delta(\tilde{\omega}_A+\omega)\delta(\omega_A+\tilde{\omega}_A-\boldsymbol{k}_A\cdot\boldsymbol{v})\frac{e^{-i\boldsymbol{k}_A\cdot\boldsymbol{b}}F_{\boldsymbol{k}_e g}^{A,0}(\boldsymbol{k}_A)F_{ge}^{B,0}(\boldsymbol{k}_A)}{\left(\frac{\tilde{\omega}_A}{c}\right)^2-k_A^2+i\eta},$$

$$I_2 = I_2^{ge,\omega} = \int d^4k_A\,\delta(\tilde{\omega}_A+\omega)\delta(\omega_A+\tilde{\omega}_A-\boldsymbol{k}_A\cdot\boldsymbol{v})\frac{e^{-i\boldsymbol{k}_A\cdot\boldsymbol{b}}\boldsymbol{F}_{\boldsymbol{k}_e g}^{A,1}(\boldsymbol{k}_A)\cdot\boldsymbol{F}_{ge}^{B,1}(\boldsymbol{k}_A)}{\left(\frac{\tilde{\omega}_A}{c}\right)^2-k_A^2+i\eta},$$

$$\boldsymbol{I}_3 = \boldsymbol{I}_3^{ge,\omega} = \int d^4k_A\,\delta(\tilde{\omega}_A+\omega)\delta(\omega_A+\tilde{\omega}_A-\boldsymbol{k}_A\cdot\boldsymbol{v})\frac{e^{-i\boldsymbol{k}_A\cdot\boldsymbol{b}}F_{\boldsymbol{k}_e g}^{A,0}(\boldsymbol{k}_A)\boldsymbol{F}_{ge}^{B,1}(\boldsymbol{k}_A)}{\left(\frac{\tilde{\omega}_A}{c}\right)^2-k_A^2+i\eta}. \tag{3.39}$$

Then, the transition amplitude for ionization of atom A via 2CPI is given by

$$a^{\pm}_{2CPI} = -\frac{ic}{\pi} D^{\pm}\left(I_1 + \frac{I_2}{4c^2} + \frac{\boldsymbol{v}\cdot\boldsymbol{I}_3}{2c^2} \right). \tag{3.40}$$

In (3.39), we may perform the integration over the frequency $\tilde{\omega}_A$ by taking advantage of the delta function $\delta(\tilde{\omega}_A + \omega)$ and rewrite the results in the following form

$$\begin{aligned}
I_1 &= -\frac{1}{c}\langle \varphi_{\boldsymbol{k}_e}(\boldsymbol{r})\chi_g(\boldsymbol{x})|I_4|\varphi_g(\boldsymbol{r})\chi_e(\boldsymbol{x})\rangle, \\
I_2 &= \frac{1}{c}\langle \varphi_{\boldsymbol{k}_e}(\boldsymbol{r})\chi_g(\boldsymbol{x})|I_4\hat{\boldsymbol{p}}_r\hat{\boldsymbol{p}}_x + \hat{\boldsymbol{p}}_x I_4\hat{\boldsymbol{p}}_r + \hat{\boldsymbol{p}}_r I_4\hat{\boldsymbol{p}}_x + \hat{\boldsymbol{p}}_r\hat{\boldsymbol{p}}_x I_4|\varphi_g(\boldsymbol{r})\chi_e(\boldsymbol{x})\rangle, \\
\boldsymbol{I}_3 &= \frac{1}{c}\langle \varphi_{\boldsymbol{k}_e}(\boldsymbol{r})\chi_g(\boldsymbol{x})|I_4\hat{\boldsymbol{p}}_x + \hat{\boldsymbol{p}}_x I_4|\varphi_g(\boldsymbol{r})\chi_e(\boldsymbol{x})\rangle,
\end{aligned} \tag{3.41}$$

where

$$I_4 = \int d^3\boldsymbol{k}_A\, \delta(\omega_A - \omega - \boldsymbol{k}_A\cdot\boldsymbol{v}) \frac{e^{-i\boldsymbol{k}_A\cdot(\boldsymbol{b}+\boldsymbol{r}-\boldsymbol{x})}}{\left(\frac{\omega}{c}\right)^2 - \boldsymbol{k}_A^2 + i\eta}. \tag{3.42}$$

Splitting the coordinates $\boldsymbol{k}_A$, $\boldsymbol{r}$ and $\boldsymbol{x}$ into their transverse parts, $\boldsymbol{k}_{A_\perp}$, $\boldsymbol{r}_\perp$ and $\boldsymbol{x}_\perp$, and longitudinal parts, $k_{A_\parallel}$, $r_\parallel$ and $x_\parallel$, respectively, as counted from the collision velocity $\boldsymbol{v}$, the integral I_4 becomes

$$I_4 = \int d^2\boldsymbol{k}_{A_\perp} \int_{-\infty}^{\infty} dk_{A_\parallel}\, \delta(\omega_A - \omega - k_{A_\parallel} v) \frac{e^{-i[\boldsymbol{k}_{A_\perp}\cdot(\boldsymbol{b}+\boldsymbol{r}_\perp-\boldsymbol{x}_\perp)+k_{A_\parallel}(r_\parallel - x_\parallel)]}}{\left(\frac{\omega}{c}\right)^2 - k_{A_\perp}^2 - k_{A_\parallel}^2 + i\eta}. \tag{3.43}$$

In (3.43), we integrate over $k_{A_\parallel}$ by applying the delta function $\delta(\omega_A - \omega - k_{A_\parallel} v) = \delta([\omega_A - \omega]/v - k_{A_\parallel})/v$ and obtain

$$I_4 = \frac{e^{-i\frac{\omega_A-\omega}{v}(r_\parallel - x_\parallel)}}{v} \int d^2\boldsymbol{k}_{A_\perp} \frac{e^{-i\boldsymbol{k}_{A_\perp}\cdot(\boldsymbol{b}+\boldsymbol{r}_\perp-\boldsymbol{x}_\perp)}}{\left(\frac{\omega}{c}\right)^2 - k_{A_\perp}^2 - \left(\frac{\omega_A-\omega}{v}\right)^2 + i\eta}. \tag{3.44}$$

Next, we introduce the vector

$$\boldsymbol{q} = \left(\boldsymbol{k}_{A_\perp}, \frac{\omega_A - \omega}{v}\right) = (\boldsymbol{q}_\perp, q_\parallel). \tag{3.45}$$

It describes the momentum transferred in the collision between atoms A and B, where $\boldsymbol{q}_\perp$ and $q_\parallel$ are the transverse and longitudinal parts (with respect to the collision velocity $\boldsymbol{v}$) of the momentum transfer, respectively. In addition, we use the notation $\boldsymbol{\rho} = \boldsymbol{b} + \boldsymbol{r}_\perp - \boldsymbol{x}_\perp$. Then, (3.44) can be written as

$$I_4 = -\frac{e^{-iq_\parallel(r_\parallel - x_\parallel)}}{v}\mathcal{J} \tag{3.46}$$

with

$$\mathcal{J} = \int d^2\boldsymbol{q}_\perp \frac{e^{-i\boldsymbol{q}_\perp\cdot\boldsymbol{\rho}}}{q_\perp^2 + q_\parallel^2 - \left(\frac{\omega}{c}\right)^2 - i\eta}. \tag{3.47}$$

In contrast to the standard theories of nonrelativistic collisions of light atomic particles, in the present calculation the electromagnetic field which transmits the interaction between A and B is not approximated by an instantaneous Coulomb form but is described relativistically that includes the presence of the retardation term $-\left(\frac{\omega}{c}\right)^2$ in the denominator of the integrand in (3.47). The retardation term allows the singularity of the integrand in (3.47) to appear at real $q_\perp$. This becomes possible if

$$q_\parallel^2 - \left(\frac{\omega}{c}\right)^2 < 0 \tag{3.48}$$

corresponding to the range of electron emission energies

$$\varepsilon_g + \omega\left(1 - \frac{v}{c}\right) < \varepsilon_{k_e} < \varepsilon_g + \omega\left(1 + \frac{v}{c}\right). \tag{3.49}$$

Note that at $v \ll 1$ a.u., this resonant energy range is quite narrow. Within and outside of it, the process of 2CPI can be considered as proceeding via different physical mechanisms. On the one hand, within the energy interval (3.49), the coupling of the $A - B$ system to the radiation field becomes efficient and 2CPI proceeds via the exchange of an on-mass-shell photon whose four-momentum $q^\mu = (\omega/c, \boldsymbol{q})$ satisfies the on-shell condition

$$q^\mu q_\mu = \left(\frac{\omega}{c}\right)^2 - \boldsymbol{q}^2 = 0 \tag{3.50}$$

which reflects the real pole of the integrand in (3.47). In the restframe of B, the on-shell photon has a frequency ω while in the restframe of A its frequency ω' is Doppler shifted and occupies the range $\omega(1 - \frac{v}{c}) < \omega' < \omega(1 + \frac{v}{c})$. On the other hand, outside the energy interval (3.49), 2CPI takes place by the exchange of an off-shell photon that is reflected by the complex pole of the integrand in (3.47).

In the following, we restrict our treatment of 2CPI to those collisions between atoms A and B for which the energies of the emitted electrons populate the range (3.49) and the interaction between A and B is exchanged by an on-shell photon. Consequently, the integral $\mathcal{J}$ in (3.47) will be evaluated under the constraint (3.48).

In (3.47), we introduce the polar coordinates $\boldsymbol{q}_\perp = (q_\perp \cos\varphi_{q_\perp}, q_\perp \sin\varphi_{q_\perp}, 0)$ and $\boldsymbol{\rho} = (\rho\cos\varphi_\rho, \rho\sin\varphi_\rho, 0)$, which yields

$$\mathcal{J} = \int_0^\infty dq_\perp \frac{q_\perp}{q_\perp^2 + q_\parallel^2 - (\frac{\omega}{c})^2 - i\eta} \int_0^{2\pi} d\varphi_{q_\perp}\, e^{-iq_\perp\rho\cos(\varphi_{q_\perp} - \varphi_\rho)}. \quad (3.51)$$

Here, the integral over the azimuthal angle $\varphi_{q_\perp}$ can be calculated straightforwardly and we arrive at

$$\mathcal{J} = 2\pi \int_0^\infty dq_\perp \frac{q_\perp J_0(q_\perp\rho)}{q_\perp^2 + q_\parallel^2 - (\frac{\omega}{c})^2 - i\eta}, \quad (3.52)$$

where $J_0(x)$ is the Bessel function [87].

Now, we use the asymptotic expansion [87]

$$J_0(q_\perp\rho) \approx \sqrt{\frac{2}{\pi q_\perp \rho}} \cos(q_\perp\rho - \frac{\pi}{4}) \quad (3.53)$$

which is valid for large arguments $q_\perp\rho \to \infty$. Since the vast majority of electrons ejected from A via 2CPI driven by on-shell photon exchange originate from very distant collisions where the absolute value b of the impact parameter is (extremely) large, the corresponding argument in such case, $q_\perp\rho = q_\perp|\boldsymbol{b} + \boldsymbol{r}_\perp - \boldsymbol{x}_\perp| \approx q_\perp b$, is assumed to be sufficiently large to apply (3.53) in good approximation (provided that $q_\perp \neq 0$). Then, (3.52) becomes

$$\mathcal{J} = \sqrt{\frac{2^3\pi}{\rho}} \int_0^\infty dq_\perp \frac{\sqrt{q_\perp}\cos(q_\perp\rho - \frac{\pi}{4})}{q_\perp^2 + q_\parallel^2 - (\frac{\omega}{c})^2 - i\eta}. \quad (3.54)$$

It is worth noting that, like the integrand in (3.52), the integrand in (3.54) approaches zero when $q_\perp \to 0$. In (3.54), we substitute $u = q_\perp\rho$ and get

$$\mathcal{J} = \sqrt{2^3\pi} \int_0^\infty du \, \frac{\sqrt{u}\cos(u - \frac{\pi}{4})}{u^2 - u_0^2 - i\tilde{\eta}} \tag{3.55}$$

with $-u_0^2 = \rho^2\left[q_\parallel^2 - \left(\frac{\omega}{c}\right)^2\right] < 0$ and $\tilde{\eta} = \rho^2\eta \to 0^+$. Applying $\cos(u - \frac{\pi}{4}) = \frac{1}{2}\left[e^{i(u-\frac{\pi}{4})} + e^{-i(u-\frac{\pi}{4})}\right]$, expression (3.55) reads

$$\mathcal{J} = \sqrt{2\pi}\left[e^{-i\frac{\pi}{4}} \int_0^\infty du \, \frac{\sqrt{u}e^{iu}}{u^2 - u_0^2 - i\tilde{\eta}} + e^{i\frac{\pi}{4}} \int_0^\infty du \, \frac{\sqrt{u}e^{-iu}}{u^2 - u_0^2 - i\tilde{\eta}}\right]. \tag{3.56}$$

In the second integral in (3.56), we substitute $w = -u$ and take into consideration that $\sqrt{-w} = \sqrt{w}e^{-i\frac{\pi}{2}}$. Subsequently, we obtain

$$\mathcal{J} = \sqrt{2\pi}\left[e^{-i\frac{\pi}{4}} \int_0^\infty du \, \frac{\sqrt{u}e^{iu}}{u^2 - u_0^2 - i\tilde{\eta}} + e^{-i\frac{\pi}{4}} \int_{-\infty}^0 dw \, \frac{\sqrt{w}e^{iw}}{w^2 - u_0^2 - i\tilde{\eta}}\right]. \tag{3.57}$$

Renaming $w = u$ in the second integral in (3.57) provides

$$\mathcal{J} = \sqrt{2\pi}e^{-i\frac{\pi}{4}} \int_{-\infty}^\infty du \, \frac{\sqrt{u}e^{iu}}{u^2 - u_0^2 - i\tilde{\eta}}. \tag{3.58}$$

The remaining integral in (3.58) can be solved by employing the Residue theorem, which yields

$$\mathcal{J} = i\sqrt{2\pi^3}e^{-i\frac{\pi}{4}} \frac{e^{i|u_0|}}{\sqrt{|u_0|}}. \tag{3.59}$$

Recalling that $-u_0^2 = \rho^2\left[q_\parallel^2 - \left(\frac{\omega}{c}\right)^2\right]$, we have $|u_0| = p\rho$ with $p = \sqrt{\left(\frac{\omega}{c}\right)^2 - q_\parallel^2} > 0$.

Then, the final result for the integral $\mathcal{J}$ can be written as

$$\mathcal{J} = i\sqrt{\frac{2\pi^3}{p}}e^{-i\frac{\pi}{4}} \frac{e^{ip\rho}}{\sqrt{\rho}}. \tag{3.60}$$

Next, we insert (3.60) into I_4 from (3.46) leading to

$$
\begin{aligned}
I_4 &= -\frac{i}{v}\sqrt{\frac{2\pi^3}{p}}e^{-i\frac{\pi}{4}}e^{-iq_\parallel(r_\parallel - x_\parallel)}\frac{e^{ip\rho}}{\sqrt{\rho}} \\
&= -\frac{i}{v}\sqrt{\frac{2\pi^3}{p}}e^{-i\frac{\pi}{4}}\left(\frac{e^{-iq_\parallel(r_\parallel - x_\parallel)}e^{ip|\boldsymbol{b}+\boldsymbol{r}_\perp-\boldsymbol{x}_\perp|}}{\sqrt{|\boldsymbol{b}+\boldsymbol{r}_\perp-\boldsymbol{x}_\perp|}}\right),
\end{aligned} \tag{3.61}
$$

where in the last line we have reinserted $\boldsymbol{\rho} = \boldsymbol{b} + \boldsymbol{r}_\perp - \boldsymbol{x}_\perp$.

Expression (3.61) is related to all kinds of multipole-interactions between the two active electrons in A and B. However, we are only interested in their strongest coupling, namely the dipole-dipole interaction. Therefore, in (3.61), we consider appropriate multipole expansions up to second order in $\boldsymbol{r} - \boldsymbol{x} = (\boldsymbol{r}_\perp - \boldsymbol{x}_\perp, r_\parallel - x_\parallel)$ in the last term in brackets. In particular, we expand the term $e^{-iq_\parallel(r_\parallel - x_\parallel)}$ around 0 up to second order in $r_\parallel - x_\parallel$ and the terms $e^{ip|\boldsymbol{b}+\boldsymbol{r}_\perp-\boldsymbol{x}_\perp|}$ and $1/\sqrt{|\boldsymbol{b}+\boldsymbol{r}_\perp-\boldsymbol{x}_\perp|}$ around $\boldsymbol{b}$ up to second order in $\boldsymbol{r}_\perp - \boldsymbol{x}_\perp$, respectively. Afterwards, we build the product of these expansions and keep only terms up to second order in $\boldsymbol{r} - \boldsymbol{x}$. Moreover, since collisions with extremely large impact parameters b ($b \gg c/\omega$) give the overwhelming contribution to the process of 2CPI in case when it is driven by the exchange of an on-shell photon, in the final expansion of (3.61), we can omit terms that are not of leading order $\sim 1/\sqrt{b}$ in b. Taking all this into account, I_4 in (3.61) is approximated by

$$
\begin{aligned}
I_4 \approx &-\frac{i}{v}\sqrt{\frac{2\pi^3}{p}}e^{-i\frac{\pi}{4}}e^{ipb}\left[\frac{1}{\sqrt{b}} + ip\frac{(\boldsymbol{r}_\perp - \boldsymbol{x}_\perp)\cdot\boldsymbol{b}}{b^{3/2}} - \frac{p^2}{2}\frac{[(\boldsymbol{r}_\perp - \boldsymbol{x}_\perp)\cdot\boldsymbol{b}]^2}{b^{5/2}} - iq_\parallel\frac{r_\parallel - x_\parallel}{\sqrt{b}}\right. \\
&\left. + q_\parallel p\frac{(r_\parallel - x_\parallel)[(\boldsymbol{r}_\perp - \boldsymbol{x}_\perp)\cdot\boldsymbol{b}]}{b^{3/2}} - \frac{q_\parallel^2}{2}\frac{(r_\parallel - x_\parallel)^2}{\sqrt{b}}\right].
\end{aligned} \tag{3.62}
$$

Now, within some elaborate but basic steps, we first insert (3.62) into I_1, I_2 and I_3 in (3.41) and subsequently keep only those matrix elements that will lead to dipole-allowed transitions in atoms A and B. Substituting the resulting expressions for I_1, I_2 and I_3 into the transition amplitude $a^{\pm}_{2CPI}$ in (3.40), the latter is obtained to be

$$
a^{\pm}_{2CPI}(\boldsymbol{b}) = \frac{1}{v}\sqrt{\frac{2\pi}{p}}e^{-i\frac{\pi}{4}}e^{ipb}D^{\pm}\Lambda(\boldsymbol{b}). \tag{3.63}
$$

Here, the quantity $\Lambda(\boldsymbol{b})$, which depends on the internal transitions of A and B and which is quite cumbersome, is given by equations (9.34)–(9.37) in Appendix 9.4 in the Electronic Supplementary Material.

In what follows, we present results for the amplitude (3.63) when φ_g and χ_g are s states. Since atom B is excited by a laser field of linear polarization, the field resonantly couples to dipole transitions where the excited state χ_e of B has a magnetic quantum number $m_B = 0$. Furthermore, we assume that $\omega_B \approx \omega$ for those terms in (3.63) which smoothly depend on ω_B. (This is a very good approximation because the laser field is resonantly coupled to B, meaning that its frequency ω lies within a very narrow interval centered at the atomic transition frequency ω_B.)

Using the set (n_j, l_j, m_j) of principal, orbital, and magnetic quantum numbers of atom j ($j = A, B$), we may separate the bound states of A and B into their radial and angular parts according to

$$\begin{aligned}\varphi_g(\boldsymbol{r}) &= \phi_{n_A,l_A=0,m_A=0}(\boldsymbol{r}) = g_{n_A}^{l_A=0}(r)Y_{l_A=0}^{m_A=0}(\vartheta_{\boldsymbol{r}}, \varphi_{\boldsymbol{r}}),\\ \chi_g(\boldsymbol{x}) &= \chi_{n_B,l_B=0,m_B=0}(\boldsymbol{x}) = h_{n_B}^{l_B=0}(x)Y_{l_B=0}^{m_B=0}(\vartheta_{\boldsymbol{x}}, \varphi_{\boldsymbol{x}}),\\ \chi_e(\boldsymbol{x}) &= \chi_{n'_B,l_B=1,m_B=0}(\boldsymbol{x}) = h_{n'_B}^{l_B=1}(x)Y_{l_B=1}^{m_B=0}(\vartheta_{\boldsymbol{x}}, \varphi_{\boldsymbol{x}})\end{aligned} \tag{3.64}$$

with $g_{n_A}^{l_A=0}$ the radial part and $Y_{l_A=0}^{m_A=0}$ the angular part of the ground state of atom A. Similarly, $h_{n_B}^{l_B=0}$ and $h_{n'_B}^{l_B=1}$ $(n_B < n'_B)$ are the radial parts and $Y_{l_B=0}^{m_B=0}$ and $Y_{l_B=1}^{m_B=0}$ the angular parts of the ground and excited state of atom B, respectively. Here, the angular parts $Y_{l_j}^{m_j}$ are described by the spherical harmonics. The continuum state of the electron emitted from A is also separated into radial and angular parts and can be expressed by

$$\varphi_{\boldsymbol{k}_e}(\boldsymbol{r}) = \frac{1}{\sqrt{V_{el}}}\frac{2\pi}{k_e}\sum_{l_A=0}^{\infty} i^{l_A}e^{-i\delta_{l_A}}g_{k_e}^{l_A}(r)\sum_{m_A=-l_A}^{l_A} Y_{l_A}^{m_A}(\vartheta_{\boldsymbol{k}_e}, \varphi_{\boldsymbol{k}_e})\big[Y_{l_A}^{m_A}(\vartheta_{\boldsymbol{r}}, \varphi_{\boldsymbol{r}})\big]^*, \tag{3.65}$$

where $g_{k_e}^{l_A}$ is the radial function of the continuum state, V_{el} the normalization volume for the electron emitted from A and $e^{-i\delta_{l_A}}$ is a phase factor. Since we are only interested in dipole-allowed bound-continuum transitions between the ground state of A with $l_A = 0$ and its continuum state with $l_A = 1$, in (3.65), only the term with $l_A = 1$ is kept.

First, we focus on the matrix element $\mathcal{W}_{ge}$ given by (3.11) which determines the quantity $D^{\pm}$ in the amplitude (3.63). Taking advantage of the commutator relation $\hat{\boldsymbol{p}}_{\boldsymbol{x}} = i[\hat{H}_B, \boldsymbol{x}]$ and accounting for the fact that χ_g and χ_e are eigenstates of the atomic Hamiltonian $\hat{H}_B$, we obtain the relation $\langle\chi_g|\hat{\boldsymbol{p}}_{\boldsymbol{x}}|\chi_e\rangle = -i\omega_B\langle\chi_g|\boldsymbol{x}|\chi_e\rangle$. Employing this relation to expression (3.11), the latter becomes

$$\mathcal{W}_{ge} = \frac{F_0}{2i} \boldsymbol{e}_z \cdot \mathcal{M}_B \tag{3.66}$$

with

$$\mathcal{M}_B = \langle \chi_g(\boldsymbol{x}) | \boldsymbol{x} | \chi_e(\boldsymbol{x}) \rangle. \tag{3.67}$$

Now, in (3.67), we apply the electronic states from (3.64) and calculate the angular integrals over $\varphi_{\boldsymbol{x}}$ and $\vartheta_{\boldsymbol{x}}$, leading to

$$\mathcal{M}_B = \frac{r_B}{\sqrt{3}} \boldsymbol{e}_z. \tag{3.68}$$

Here, $r_B = \int_0^\infty dx\, x^3 \big[h_{n_B}^{l_B=0}(x)\big]^* h_{n_B'}^{l_B=1}(x)$ is the radial matrix element for transitions from the excited state χ_e into the ground state χ_g in B. Insertion of (3.68) into (3.66) provides

$$\mathcal{W}_{ge} = \frac{F_0 r_B}{\sqrt{12} i}. \tag{3.69}$$

Next, we consider the quantity $\Lambda(\boldsymbol{b})$ which enters the amplitude (3.63) and which is determined by equations (9.34)–(9.37) in Appendix 9.4 in the Electronic Supplementary Material. Employing the electronic states from (3.64) and (3.65), the quantity $\Lambda(\boldsymbol{b})$ is obtained to be (see Appendix 9.4 in the Electronic Supplementary Material)

$$\Lambda(\boldsymbol{b}) = \sqrt{\frac{\pi}{3}} \frac{e^{i\delta_1}}{i} \frac{r_A r_B}{\sqrt{V_{el}} k_e^2} \frac{\big[q_\parallel^2 - \big(\frac{\omega}{c}\big)^2\big] k_{e_\parallel} - q_\parallel p \cos(\varphi_b - \varphi_{k_{e_\perp}}) k_{e_\perp}}{\sqrt{b}}. \tag{3.70}$$

Inserting (3.70) into (3.63), the transition amplitude for 2CPI via the coupling to the radiation field can be written as

$$a_{2CPI}^{\pm}(\boldsymbol{b}) = e^{i\alpha} \sqrt{\frac{2\pi^2}{3}} \frac{D^{\pm} r_A r_B}{v \sqrt{V_{el}} p k_e^2} \frac{\big[q_\parallel^2 - \big(\frac{\omega}{c}\big)^2\big] k_{e_\parallel} - q_\parallel p \cos(\varphi_b - \varphi_{k_{e_\perp}}) k_{e_\perp}}{\sqrt{b}}, \tag{3.71}$$

where $\alpha = pb + \delta_1 - \frac{3\pi}{4}$. Note that the amplitude in (3.71) scales as $1/\sqrt{b}$ and therefore the transition probability $|a_{2CPI}^{\pm}|^2$ behaves as $1/b$. This reflects the very long range of the interatomic interaction in case when atoms A and B interact with each other by the exchange of an on-shell photon.

3.1.5 Cross Section and Reaction Rate for Two-Center Photoionization Via Coupling to the Radiation Field

The spectrum of electrons emitted from atom A by the process of 2CPI via the coupling to the radiation field is determined by the cross section differential in the electron momentum $\boldsymbol{k}_e$, which is given by

$$\frac{d^3\sigma^{\pm}_{2CPI}}{d\boldsymbol{k}_e^3} = \frac{V_{el}}{(2\pi)^3}\int d^2\boldsymbol{b}\,|a^{\pm}_{2CPI}(\boldsymbol{b})|^2 = \frac{V_{el}}{(2\pi)^3}\int_0^{2\pi} d\varphi_{\boldsymbol{b}}\int_{b_1}^{b_{max}} db\, b\,|a^{\pm}_{2CPI}(\boldsymbol{b})|^2. \tag{3.72}$$

In (3.72), the integrations run over the azimuthal angle $\varphi_{\boldsymbol{b}}$ and the absolute value b of the impact parameter $\boldsymbol{b}$. The integration for the absolute value of the impact parameter is limited to the range between $b = b_1$ and $b = b_{max}$. Here, b_{max} is the maximum possible value of b in the collision and b_1 is not too small in order to justify the usage of the asymptotic expansion (3.53) for the Bessel function in the derivation of the transition amplitude $a^{\pm}_{2CPI}$. We mention that in experiments, b_{max} is typically of macroscopic size ($b_{max} \sim 1 - 10$ mm) and thus one has $b_{max} \gg b_1$.

Substituting the amplitude (3.71) into (3.72), we get

$$\frac{d^3\sigma^{\pm}_{2CPI}}{d\boldsymbol{k}_e^3} = \frac{1}{12\pi}\frac{|D^{\pm}|^2 r_A^2 r_B^2}{v^2 p k_e^4}\cdot \times\left\{\left[q_{\parallel}^2 - \left(\frac{\omega}{c}\right)^2\right]^2 k_{e_\parallel}^2 I_8 - 2\left[q_{\parallel}^2 - \left(\frac{\omega}{c}\right)^2\right] q_{\parallel} p k_{e_\parallel} k_{e_\perp} I_9 + q_{\parallel}^2 p^2 k_{e_\perp}^2 I_{10}\right\} \tag{3.73}$$

with

$$I_8 = \int_0^{2\pi} d\varphi_{\boldsymbol{b}}\int_{b_1}^{b_{max}} db,$$
$$I_9 = \int_0^{2\pi} d\varphi_{\boldsymbol{b}}\,\cos(\varphi_{\boldsymbol{b}} - \varphi_{\boldsymbol{k}_{e_\perp}})\int_{b_1}^{b_{max}} db,$$
$$I_{10} = \int_0^{2\pi} d\varphi_{\boldsymbol{b}}\,\cos^2(\varphi_{\boldsymbol{b}} - \varphi_{\boldsymbol{k}_{e_\perp}})\int_{b_1}^{b_{max}} db. \tag{3.74}$$

The integrals in (3.74) are easily calculated and using $b_{max} \gg b_1$ their results read

$$I_8 = 2\pi b_{max},$$
$$I_9 = 0,$$
$$I_{10} = \pi b_{max}. \tag{3.75}$$

Insertion of (3.75) into (3.73) yields the cross section

$$\frac{d^3\sigma^{\pm}_{2CPI}}{d\boldsymbol{k}_e^3} = \frac{1}{12}\frac{|D^{\pm}|^2 r_A^2 r_B^2 b_{max}}{v^2 p k_e^4}\left\{2\left[q_{\parallel}^2 - \left(\frac{\omega}{c}\right)^2\right]^2 k_{e_{\parallel}}^2 + q_{\parallel}^2 p^2 k_{e_{\perp}}^2\right\}. \tag{3.76}$$

Expression (3.76) contains the quantities r_A^2 and r_B^2 which can be expressed via the cross section σ_{PI}^A for direct photoionization of atom A and the radiative width Γ_r^B of the excited state in atom B, respectively, according to

$$r_A^2 = \frac{3ck_e}{2\pi\omega}\sigma_{PI}^A(\omega),$$
$$r_B^2 = \frac{9c^3}{4\omega^3}\Gamma_r^B. \tag{3.77}$$

Then, substituting (3.77) into (3.76), the cross section differential in the electron momentum for 2CPI via the coupling to the radiation field can be written as

$$\frac{d^3\sigma^{\pm}_{2CPI}}{d\boldsymbol{k}_e^3} = \frac{9}{32\pi}\frac{|D^{\pm}|^2 \sigma_{PI}^A(\omega)\Gamma_r^B b_{max}}{v^2 p k_e^3}\left(\frac{c}{\omega}\right)^4\left\{2\left[q_{\parallel}^2 - \left(\frac{\omega}{c}\right)^2\right]^2 k_{e_{\parallel}}^2 + q_{\parallel}^2 p^2 k_{e_{\perp}}^2\right\}. \tag{3.78}$$

The cross section differential in the solid angle $\Omega_{\boldsymbol{k}_e}$ for 2CPI when it proceeds via the exchange of an on-shell photon is given by

$$\frac{d^2\sigma^{\pm}_{2CPI}}{d\Omega_{\boldsymbol{k}_e}} = \int_{q_{\parallel}^2 < (\frac{\omega}{c})^2} dk_e\, k_e^2\, \frac{d^3\sigma^{\pm}_{2CPI}}{d\boldsymbol{k}_e^3}. \tag{3.79}$$

Here, we integrate over the absolute value k_e of the electron momentum $\boldsymbol{k}_e$ where the integration interval is determined by the condition $q_{\parallel}^2 < \left(\frac{\omega}{c}\right)^2$ (for which 2CPI occurs via the exchange of an on-shell photon).

Now, we insert (3.78) into (3.79). Moreover, we introduce the spherical coordinates $\boldsymbol{k}_e = (k_e \sin\vartheta_{\boldsymbol{k}_e}\cos\varphi_{\boldsymbol{k}_e},\ k_e \sin\vartheta_{\boldsymbol{k}_e}\sin\varphi_{\boldsymbol{k}_e},\ k_e\cos\vartheta_{\boldsymbol{k}_e})$, such that $k_{e\perp}^2 = k_{e_x}^2 + k_{e_y}^2 = k_e^2\sin^2\vartheta_{\boldsymbol{k}_e}$ and $k_{e_{\parallel}}^2 = k_{e_z}^2 = k_e^2\cos^2\vartheta_{\boldsymbol{k}_e}$. Afterwards, we arrive at

$$\frac{d^2\sigma^{\pm}_{2CPI}}{d\Omega_{k_e}} = \frac{9}{32\pi}\frac{|D^{\pm}|^2\sigma^A_{PI}(\omega)\Gamma^B_r b_{max}}{v^2}\left(\frac{c}{\omega}\right)^4\left\{2\cos^2\vartheta_{k_e}I_{11} + \sin^2\vartheta_{k_e}I_{12}\right\} \quad (3.80)$$

with

$$\begin{aligned} I_{11} &= \int_{q_\parallel^2<(\frac{\omega}{c})^2} dk_e\,\frac{k_e}{p}\left[q_\parallel^2 - \left(\frac{\omega}{c}\right)^2\right]^2, \\ I_{12} &= \int_{q_\parallel^2<(\frac{\omega}{c})^2} dk_e\,k_e p q_\parallel^2. \end{aligned} \quad (3.81)$$

We remind that $p = \sqrt{\left(\frac{\omega}{c}\right)^2 - q_\parallel^2}$ and rewrite the integrals over the coordinate k_e in (3.81) into integrals over the coordinate $q_\parallel$ according to

$$\begin{aligned} I_{11} &= v\left(\frac{\omega}{c}\right)^3\int_{-\omega/c}^{\omega/c} dq_\parallel\,\frac{\left[1-\left(\frac{c}{\omega}q_\parallel\right)^2\right]^2}{\sqrt{1-\left(\frac{c}{\omega}q_\parallel\right)^2}}, \\ I_{12} &= v\left(\frac{\omega}{c}\right)^3\int_{-\omega/c}^{\omega/c} dq_\parallel\left(\frac{c}{\omega}q_\parallel\right)^2\sqrt{1-\left(\frac{c}{\omega}q_\parallel\right)^2}. \end{aligned} \quad (3.82)$$

Performing the substitution $u = \frac{c}{\omega}q_\parallel$ in (3.82) provides

$$\begin{aligned} I_{11} &= v\left(\frac{\omega}{c}\right)^4\int_{-1}^{1} du\,\frac{[1-u^2]^2}{\sqrt{1-u^2}} = 2v\left(\frac{\omega}{c}\right)^4\int_0^1 du\,\frac{[1-u^2]^2}{\sqrt{1-u^2}}, \\ I_{12} &= v\left(\frac{\omega}{c}\right)^4\int_{-1}^{1} du\,u^2\sqrt{1-u^2} = 2v\left(\frac{\omega}{c}\right)^4\int_0^1 du\,u^2\sqrt{1-u^2}. \end{aligned} \quad (3.83)$$

The solutions of the basic integrals in (3.83) read

$$\begin{aligned} \int_0^1 du\,\frac{[1-u^2]^2}{\sqrt{1-u^2}} &= \frac{3\pi}{16}, \\ \int_0^1 du\,u^2\sqrt{1-u^2} &= \frac{\pi}{16} \end{aligned} \quad (3.84)$$

and consequently I_{11} and I_{12} are obtained to be

$$I_{11} = \frac{3\pi}{8} v \left(\frac{\omega}{c}\right)^4,$$

$$I_{12} = \frac{\pi}{8} v \left(\frac{\omega}{c}\right)^4. \tag{3.85}$$

Inserting (3.85) into (3.80), the cross section differential in the solid angle Ω_{k_e} for 2CPI when it proceeds via the exchange of an on-shell photon becomes

$$\frac{d^2\sigma^{\pm}_{2CPI}}{d\Omega_{k_e}} = \frac{9}{256} |D^{\pm}|^2 \sigma^A_{PI}(\omega) \Gamma^B_r \frac{b_{max}}{v} \left\{1 + 5\cos^2\vartheta_{k_e}\right\}. \tag{3.86}$$

The total cross section for 2CPI via the coupling to the radiation field is given by

$$\begin{aligned} \sigma^{\pm}_{2CPI} &= \int d\Omega_{k_e} \frac{d^2\sigma^{\pm}_{2CPI}}{d\Omega_{k_e}} \\ &= \int_0^{2\pi} d\varphi_{k_e} \int_0^{\pi} d\vartheta_{k_e} \sin\vartheta_{k_e} \frac{d^2\sigma^{\pm}_{2CPI}}{d\Omega_{k_e}}, \end{aligned} \tag{3.87}$$

where we integrate over the angles φ_{k_e} and ϑ_{k_e} of the electron momentum $\boldsymbol{k}_e$.

Substituting (3.86) into (3.87) and calculating the angular integrals, the total cross section can be written as

$$\sigma^{\pm}_{2CPI} = |D^{\pm}|^2 \times \frac{3\pi}{8} \frac{\sigma^A_{PI}(\omega) \Gamma^B_r b_{max}}{v}. \tag{3.88}$$

For the sake of completeness, we note that the quantity $|D^{\pm}|^2$ in (3.88) is determined by $|W_{ge}|^2$ which, using (3.69) and (3.77), reads $|W_{ge}|^2 = 3c^3 F_0^2 \Gamma^B_r/(16\omega^3)$.

It is worth mentioning that the cross section in (3.88) has a rather simple structure that allows for a straightforward interpretation of the process of 2CPI via the coupling to the radiation field. The factor $|D^{\pm}|^2$ describes the probability for atom B to enter the collision in an excited state and the remaining part $\frac{3\pi}{8} \frac{\sigma^A_{PI}(\omega)\Gamma^B_r b_{max}}{v}$ represents the cross section for ionization of atom A in collisions with excited atom B by the exchange of an on-shell photon between them. The magnitude of the numerical prefactor in the latter cross section depends on the specific dipole-allowed transition which is involved in de-excitation of B during the collision.

Let us suppose for the moment that the interatomic distance R between A and B is fixed and that this distance is very large, $R \gg c/\omega$. Further, we only consider the second step of 2CPI in which atom B de-excites and the de-excitation energy is transferred to atom A leading to its ionization. In such case, the relativistic form of the dipole-dipole interaction $\hat{V}_{AB}$ between atoms A and B is given by equation (2.4). It accounts for the retardation effect and therefore implies the coupling of the $A-B$ system to the radiation field. We may use $\hat{V}_{AB}$ in order to calculate the decay rate Γ_1 for the second step of 2CPI when the distance between A and B is fixed, yielding $\Gamma_1 \propto \frac{\sigma_{PI}^A(\omega)\Gamma_r^B}{R^2}$. Now, we can obtain the corresponding cross section σ_1 for collisions between atoms A and B by introducing $\boldsymbol{R}(t) = \boldsymbol{b} + \boldsymbol{v}t$ and subsequently performing the integration $\sigma_1 = \int d^2\boldsymbol{b} \int_{-\infty}^{\infty} dt\ \Gamma_1(\boldsymbol{b}, t)$ under the assumption $b \in [b_1, b_{max}]$. The resulting cross section is proportional to $\frac{\sigma_{PI}^A(\omega)\Gamma_r^B b_{max}}{v}$. The latter term also appears in the cross section (3.88), so based on the above simple consideration, we can conclude that this term arises because in our approach to the process of 2CPI the ionization proceeds via the coupling to the radiation field.

Taking into account that atom A moves in a gas of atoms B having an atomic density n_B, the reaction rate per unit of time (per atom A) corresponding to the total cross section (3.88) is obtained to be

$$\mathcal{R}_{2CPI}^{\pm} = \sigma_{2CPI}^{\pm} n_B v = |D^{\pm}|^2 \times \frac{3\pi}{8} \sigma_{PI}^A(\omega)\Gamma_r^B b_{max} n_B. \qquad (3.89)$$

3.2 Numerical Results and Discussion

In this Section, we present numerical results of our theoretical findings in Section 3.1 for the relativistic two-center photoionization channel (in which the interaction between atoms A and B incorporates retardation) and compare them to the results of the nonrelativistic two-center channel (where the interatomic interaction is considered in its instantaneous Coulomb form). Further, we discuss the relative effectiveness of 2CPI compared with direct photoionization of atom A.

3.2.1 Two-Center Photoionization Via Two-Center Autoionization

In the theoretical consideration in Section 3.1, we have pointed out that the process of 2CPI occurring in distant collisions between atoms A and B can be splitted into two contributions. The relativistic two-center channel, for which results were obtained in Section 3.1, describes those collisions where 2CPI proceeds via the exchange of an on-shell photon between B and A resulting in electron emission into the continuum of A with energies $\varepsilon_g + \omega(1 - v/c) < \varepsilon_{k_e} < \varepsilon_g + \omega(1 + v/c)$.

In contrast, the nonrelativistic channel takes into account collisions for which 2CPI involves the exchange of an off-shell photon between B and A with consequent electron emission into the continuum of A having energies outside the above energy range.

The cross section $\sigma^{\pm}_{2CPI,nr}$ for 2CPI via the exchange of an off-shell photon was obtained in [49] for a laser field of linear polarization ($\boldsymbol{F}_0 \parallel \boldsymbol{v}$) using the first order of perturbation theory in the interaction between atom B and the laser field[1]. It can be expressed in the form

$$\sigma^{\pm}_{2CPI,nr} = |\beta^{\pm}|^2 \times \frac{9\pi}{64}\left(\frac{c}{\omega}\right)^4 \frac{\sigma^{A}_{PI}(\omega)\Gamma^{B}_{r}}{v b_0^3}. \tag{3.90}$$

Here, $|\beta^{+}|^2 = |\mathcal{W}_{ge}|^2/(\Delta^2 + (\Gamma^B_r/2)^2)$, $|\beta^{-}|^2 \approx 0$ and b_0 is the minimum value of the impact parameter for which the electronic shells of atoms A and B do not overlap.

Similar to the cross section (3.88) for 2CPI via the exchange of an on-shell photon, the cross section in (3.90) has a simple and transparent structure. The factor $|\beta^{\pm}|^2$ represents the probability for atom B to enter the collision in an excited state and the remaining part $\frac{9\pi}{64}\left(\frac{c}{\omega}\right)^4 \frac{\sigma^{A}_{PI}(\omega)\Gamma^{B}_{r}}{v b_0^3}$ describes the cross section for ionization of atom A in collisions with excited atom B by the exchange of an off-shell photon between them. The magnitude of the numerical prefactor in the latter cross section depends on the specific dipole-allowed transition which is involved in de-excitation of B upon collisions.

Let us suppose for the moment that the interatomic distance R between A and B is fixed with $1 \ll R \ll c/\omega$. In addition, we only consider the second step of 2CPI where atom B de-excites and the de-excitation energy is transferred to atom A resulting in its ionization. In such case, the dipole-dipole interaction $\hat{V}_{AB}$ between atoms A and B can be approximated by the instantaneous interaction between two electric dipoles given by equation (1.6). We may use $\hat{V}_{AB}$ in order to calculate the decay rate Γ_2 for the second step of 2CPI when the distance between A and B is fixed. Then, we arrive at $\Gamma_2 \propto \left(\frac{c}{\omega}\right)^4 \frac{\sigma^{A}_{PI}(\omega)\Gamma^{B}_{r}}{R^6}$, which represents the known result (see, e.g. [26, 88]) for the two-center autoionization rate at large interatomic distances ($R \gg 1$ a.u.). Next, we can obtain the corresponding cross section σ_2 for collisions between atoms A and B by introducing $\boldsymbol{R}(t) = \boldsymbol{b} + \boldsymbol{v}t$ and afterwards

[1] We note that the field-dressed states of atom B were calculated by supposing that the laser field is switched on adiabatically at $t \to -\infty$ and imposing the boundary conditions which were also used in the present work (see below equation (3.9)).

calculating the integral $\sigma_2 = \int d^2\boldsymbol{b} \int_{-\infty}^{\infty} dt\ \Gamma_2(\boldsymbol{b}, t)$ under the assumption $b \geq b_0$. The resulting cross section is proportional to $\left(\frac{c}{\omega}\right)^4 \frac{\sigma_{PI}^A(\omega)\Gamma_r^B}{v b_0^3}$. The latter term also appears in the cross section (3.90), so the above simple consideration suggests that this term arises because in the nonrelativistic approach to the process of 2CPI the ionization occurs via the two-center autoionization mechanism (in which ionization of A is a consequence of the transfer of de-excitation energy from B by the exchange of an off-shell photon between B and A).

When atom A moves in a gas of atoms B with an atomic density n_B, the reaction rate per unit of time (per atom A) corresponding to the total cross section (3.90) is given by

$$\mathcal{R}^{\pm}_{2CPI,nr} = \sigma^{\pm}_{2CPI,nr} n_B v = |\beta^{\pm}|^2 \times \frac{9\pi}{64}\left(\frac{c}{\omega}\right)^4 \frac{\sigma_{PI}^A(\omega)\Gamma_r^B n_B}{b_0^3}. \qquad (3.91)$$

3.2.2 Two-Center Photoionization Via Coupling to the Radiation Field vs. Two-Center Photoionization Via Two-Center Autoionization

The relative effectiveness of the relativistic and nonrelativistic two-center photoionization channels can be characterized by the ratio of their reaction rates. In the derivation of the rate $\mathcal{R}^{\pm}_{2CPI,nr}$ for 2CPI via the exchange of an off-shell photon, the field-dressed states of atom B were calculated by using the first order of perturbation theory in the interaction between atom B and the laser field. However, in our derivation of the rate $\mathcal{R}^{\pm}_{2CPI}$ for 2CPI via the exchange of an on-shell photon, the field-dressed states of B were obtained by applying the rotating wave approximation. The latter allows the consideration of stronger fields as compared with the first order of perturbation theory. The different approaches to the calculation of the field-dressed states are manifested in the rates (3.89) and (3.91) by the quantities $|D^{\pm}|^2$ and $|\beta^{\pm}|^2$, respectively, both of which describe the probability for B to enter the collision in an excited state.

In order to get comparable results for the rates (3.89) and (3.91), we restrict our result for $\mathcal{R}^{\pm}_{2CPI}$ to the case of a weak laser field that obeys the condition $|\mathcal{W}_{ge}| \ll |\Delta|$ for which the first order of perturbation theory in the interaction between B and the laser field is applicable. Then, one can show in some basic steps that $|D^{+}|^2 \approx |\mathcal{W}_{ge}|^2/(\Delta^2 + (\Gamma_r^B/2)^2) = |\beta^{+}|^2$ and $|D^{-}|^2 \approx 0 \approx |\beta^{-}|^2$ in very good approximation. In the above case, the rate (3.89) for 2CPI via the exchange of an on-shell photon becomes

$$\mathcal{R}^{\pm}_{2CPI,r} = |\beta^{\pm}|^2 \times \frac{3\pi}{8}\sigma^A_{PI}(\omega)\Gamma^B_r b_{max} n_B. \tag{3.92}$$

In the following, we only consider the non-zero reaction rates $\mathcal{R}^{+}_{2CPI,r}$ and $\mathcal{R}^{+}_{2CPI,nr}$ in (3.92) and (3.91), respectively. The corresponding ratio of these rates takes the rather simple form

$$\eta = \frac{\mathcal{R}^{+}_{2CPI,r}}{\mathcal{R}^{+}_{2CPI,nr}} = \frac{8}{3}\left(\frac{\omega}{c}\right)^4 b_{max} b_0^3. \tag{3.93}$$

The ratio in (3.93) strongly depends on the amount of energy $\omega_B \approx \omega$ that is transferred from atom B to A. In particular, 2CPI in slow collisions involving a large (small) energy transfer will likely be dominated by the exchange of an on-shell (off-shell) photon. It is worth mentioning that the correspondence between 2CPI via the coupling to the radiation field and 2CPI via two-center autoionization closely resembles that between spontaneous radiative decay and autoionization in single atoms and ions where the former dominates transitions with a large energy release and the latter dominates transitions with a small energy release.

Let us consider two examples for 2CPI in slow collisions where it involves a large and small energy transfer, respectively, from B to A. In the first example, atomic species B is represented by He atoms which are driven by a laser field whose frequency ω is resonantly tuned to the $1\,^1S_0 - 2\,^1P_1$ dipole transition ($\omega_B \approx 21.2$ eV) in He. In the second example, we take Rb atoms as atomic species B and the laser field shall be resonant to the $5s_{1/2} - 5p_{3/2}$ dipole transition ($\omega_B \approx 1.59$ eV) in Rb. For the above examples, Fig. 3.2 shows the ratio η as a function of the maximum possible value b_{max} of the impact parameter b. We can conclude from Fig. 3.2 that when 2CPI involves the $1\,^1S_0 - 2\,^1P_1$ transition in He (and the energy transfer is large), it will be dominated by the exchange of an on-shell photon provided $b_{max} \gtrsim 1$ mm. However, when 2CPI involves the $5s_{1/2} - 5p_{3/2}$ transition in Rb (and the energy transfer is small), it will be dominated by the exchange of an off-shell photon for any realistic choice of b_{max}.

3.2.3 Two-Center Photoionization vs. Direct Photoionization

Two-center resonant photoionization competes with direct photoionization of atom A by the laser field. The reaction rate $\mathcal{R}_{DPI}$ for direct photoionization of A per unit of time (per atom A) was obtained in [49] and can be written in the form

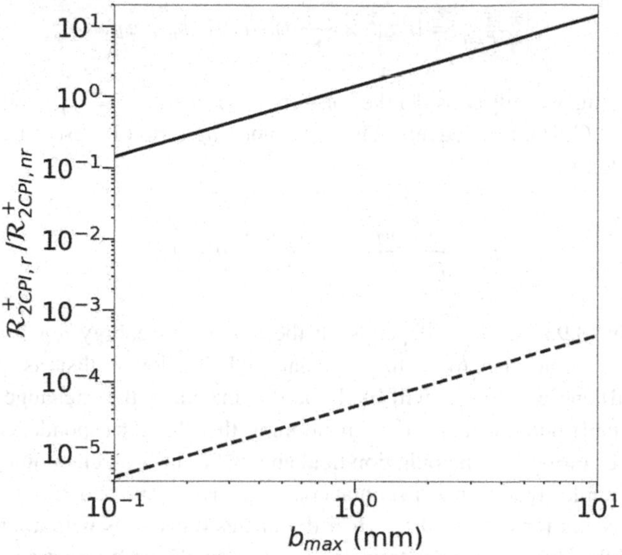

Fig. 3.2 The ratio $\eta = \mathcal{R}^+_{2CPI,r}/\mathcal{R}^+_{2CPI,nr}$ considered as a function of the maximum impact parameter b_{max} at $b_0 = 3$ a. u. for 2CPI in slow collisions involving the $1\,^1S_0 - 2\,^1P_1$ transition in He (solid) and the $5s_{1/2} - 5p_{3/2}$ transition in Rb (dashed), respectively

$$\mathcal{R}_{DPI} = \frac{cF_0^2\sigma^A_{PI}(\omega)}{8\pi\omega}. \tag{3.94}$$

The relative effectiveness of two-center and direct photoionization can be characterized by the ratio of the total rate $\mathcal{R}_{2CPI} = \mathcal{R}^+_{2CPI,r} + \mathcal{R}^+_{2CPI,nr}$ for 2CPI and the rate $\mathcal{R}_{DPI}$ for direct photoionization. Using the corresponding rates given in (3.91), (3.92) and (3.94), this ratio becomes

$$\begin{aligned}\zeta &= \frac{\mathcal{R}_{2CPI}}{\mathcal{R}_{DPI}} = \frac{\mathcal{R}^+_{2CPI,r}}{\mathcal{R}_{DPI}} + \frac{\mathcal{R}^+_{2CPI,nr}}{\mathcal{R}_{DPI}} \\ &= \frac{3\pi^2|\beta^+|^2\omega\Gamma_r^B b_{max} n_B}{cF_0^2} + \frac{9\pi^2|\beta^+|^2c^3\Gamma_r^B n_B}{8\omega^3F_0^2b_0^3} \\ &= \frac{3\pi}{8}\frac{b_{max}}{\Lambda^B_{rad}} + \frac{9\pi}{64}\frac{(c/\omega)^4}{\Lambda^B_{rad}b_0^3},\end{aligned} \tag{3.95}$$

where for the last line in (3.95) we have taken advantage of the explicit expression $|\beta^+|^2 = 3c^3 F_0^2 \Gamma_r^B/(16\omega^3(\Delta^2 + (\Gamma_r^B/2)^2))$ and afterwards introduced the mean free path $\Lambda_{rad}^B = 1/(n_B \sigma_{sc}^B)$ of the radiation in the gas of atoms B with $\sigma_{sc}^B = 3\pi/2(c/\omega)^2(\Gamma_r^B)^2/(\Delta^2 + (\Gamma_r^B/2)^2)$ denoting the cross section for resonant photon scattering on B.

The second term of the sum in (3.95) suggests that 2CPI via the exchange of an off-shell photon can be more efficient than direct photoionization as long as the energy transfer $\omega_B \approx \omega$ from B to A is sufficiently small. In the opposite case, when the energy transfer is relatively large and 2CPI is already dominated by the exchange of an on-shell photon, the ratio ζ will be determined by the first term of the sum in (3.95), involving the ratio b_{max}/Λ_{rad}^B of the maximum impact parameter b_{max} and the mean free path Λ_{rad}^B of the radiation in the gas of atoms B. Here, the magnitude of b_{max} depends on the size of the target gas of atoms B and/or the size of the projectile beam of atoms A. It can however not considerably exceed Λ_{rad}^B since otherwise the gas of atoms B becomes intransparent for the radiation with respect to the mechanism of 2CPI via the exchange of an on-shell photon. In case that $b_{max} = \Lambda_{rad}^B$, we get $\zeta = 3\pi/8$, meaning that 2CPI when it proceeds via the coupling to the radiation field can more than double the rate for ionization of atoms A.

3.2.4 Experiments on Two-Center Photoionization

We now consider a possible experiment on 2CPI which involves the simplest atoms, namely H and He. Let a beam of slow H atoms penetrate a gas of cold He atoms irradiated by a weak (intensity $I \lesssim 10^2$ W/cm^2, where $I = cF_0^2/(8\pi)$) monochromatic laser field of linear polarization ($\boldsymbol{F}_0 \parallel \boldsymbol{v}$) whose frequency is resonantly tuned to the $1\,^1S_0 - 2\,^1P_1$ transition in He ($\omega \approx 21.2$ eV). Further, we set $b_{max} = 5$ mm and $b_0 = 3$ a. u. (which is a quite conservative estimate for b_0). Then, equation (3.93) yields $\eta \approx 7$ from which we can conclude that 2CPI is strongly dominated by the exchange of an on-shell photon.

Moreover, the total cross section for photon scattering on He evaluated at the exact resonance ($\Delta = 0$) is given by $\sigma_{sc}^B = 6\pi(c/\omega)^2 \approx 1.63 \times 10^{-11}$ cm^2. Consequently, the mean free path Λ_{rad}^B of the radiation in the He gas will be equal to $b_{max} = 5$ mm at the gas density of $n_B = 1/(b_{max}\sigma_{sc}^B) \approx 1.22 \times 10^{11}$ cm^{-3}. Under the above conditions, equation (3.95) provides $\zeta = 3\pi/8$, implying that 2CPI when it takes place via the coupling to the radiation field results in a slightly larger ionization rate for H atoms than direct photoionization.

Note that in order to achieve an efficient excitation of the target atoms by the resonant laser field, the Doppler broadening of the spectral line for the $1\,^1S_0 - 2\,^1P_1$ transition in He, caused by the thermal motion of He atoms in the target gas, should be sufficiently small. By cooling down the He target to temperatures $T \lesssim 0.01$ K, the Doppler broadening becomes considerably smaller than the natural linewidth of the $2\,^1P_1 \to 1\,^1S_0$ transition in He and the laser field can efficiently excite the target atoms.

Finally, we want to point out that one can also envisage experiments in which 2CPI via on-shell photon exchange is the only ionization mechanism to occur. Here, a projectile beam of atoms A passes close by the target gas of atoms B, the latter of which is irradiated by a weak laser field. The projectile beam does not penetrate the target gas and is not exposed to the laser field. In such case, the (relative) short-range interaction between A and B via the exchange of an off-shell photon as well as the interaction of A with the laser field are excluded and only the long-range interaction between A and B via the exchange of an on-shell photon can contribute to the ionization of atom A.

3.2.5 Two-Center Photoionization and the Breit Interaction

The interaction between slowly moving charged particles usually occurs primarily via the (instantaneous) Coulomb interaction related to the exchange of time-like and longitudinal photons between them. If the interacting particles are electrons, the leading relativistic correction to the Coulomb interaction is provided by the (generalized) Breit interaction (see, e.g. [89] and references therein). The latter follows from Quantum Electrodynamics in first-order perturbation theory, where it occurs as a result of the exchange of single transverse photons between the electrons.

In our case, the contributions of the Coulomb and Breit interactions to the amplitude for 2CPI can be made explicit by using the conservation of electric charge and proceeding similarly as in [86] in order to rewrite the amplitude $a^{\pm}_{2CPI}(\boldsymbol{b})$ given by equations (3.40)–(3.41) & (3.46)–(3.47) as a sum of two terms $a^{\pm}_{2CPI}(\boldsymbol{b}) = a^{\pm}_{Coul}(\boldsymbol{b}) + a^{\pm}_{Breit}(\boldsymbol{b})$. The term $a^{\pm}_{Coul}$, whose integrand is proportional to $\boldsymbol{q}^{-2}$ which does not possess a pole on the real axis of $q_{\perp}$, refers to the exchange of time-like and longitudinal photons that transmit the Coulomb interaction between the active electrons in atoms A and B, where these photons satisfy the off-mass shell condition. The term $a^{\pm}_{Breit}$, whose integrand is proportional to $(\boldsymbol{q}^2 - \omega^2/c^2 - i\eta)^{-1}$, arises due to the exchange of transverse photons which are responsible for the Breit interaction between the electrons in A and B, where these photons can satisfy the on-mass shell or off-mass shell condition. Under the constraint $q_{\parallel}^2 - \omega^2/c^2 < 0$, corresponding to electron emission into the narrow resonant energy range $\varepsilon_g + \omega(1 - v/c) < \varepsilon_{k_e} < \varepsilon_g + \omega(1 + v/c)$, the term $(q^2 - \omega^2/c^2 - i\eta)^{-1}$ exhibits a pole at real $q_{\perp}$ and the Breit interaction is transmitted by photons that satisfy the on-mass shell condition. In such case, $a^{\pm}_{Breit}$ simply refers to the process of 2CPI when it is driven by the exchange of on-shell photons and the Breit interaction can become very efficient, even dominating over the Coulomb interaction.

This efficiency of the Breit interaction at very low energies is quite remarkable, because normally the Breit interaction plays only a relatively minor role in atomic physics, including processes in which high energy electrons are involved.

3.2.6 Ionization at Larger Collision Velocities

In our treatment of atomic collisions, we have assumed that the collision velocity v is much smaller than the typical orbiting velocities $v_e \sim 1$ a. u. of the active electrons in atoms A and B. The main reason for this assumption was that for $v \ll 1$ a. u. the impact excitation or ionization of atom A (or B) is strongly suppressed and we are effectively dealing only with direct and two-center photoionization of A (see also Section 3.1.2).

However, in the derivation of the total cross section (3.88) and the rate (3.89) resp. (3.92) for 2CPI via the coupling to the radiation field, the assumption $v \ll 1$ a. u. was not required, meaning that these results remain valid also for much larger collision velocities as long as v is much less than the speed of light c. In particular, this suggests that the rate for 2CPI when it occurs via the exchange of an on-shell photon does not depend on the collision velocity up to impact energies ~ 10 MeV/u. On the other hand, for impact energies $\gtrsim 0.5$ MeV/u, the strength of all ionization mechanisms proceeding via the exchange of off-shell photons rapidly decreases with increasing the collision velocity (see, e.g. [82]). Therefore, it is expected that the coupling of the $A - B$ system to the radiation field is also very important for ionization of atoms A at larger collision velocities.

3.3 Summary and Concluding Remarks

We have considered single electron emission from atomic species A in slow distant collisions with another species B excited by a weak laser field when it proceeds via two-center resonant photoionization driven by the coupling of the colliding system to the radiation field.

In systems in which the ionization potential of atom A is smaller than an excitation energy for a dipole-allowed transition in atom B, two-center ionization of A occurs via resonant photoexcitation of thc dipole transition in B by the laser field with subsequent de-excitation of B, where the energy excess is transferred—via the (long-range) interatomic interaction—to A, resulting in its ionization.

The theoretical treatment of collisions between A and B was based on the semiclassical approximation, in which the relative motion of the (heavy) nuclei is described classically while the active electrons are treated quantum mechanically. This approximation is well justified starting with quite low impact energies (~ 1 eV/u). We have performed a relativistic calculation which incorporates the retardation effect accounting for the finite propagation of the electromagnetic field that transmits the interaction between A and B.

2CPI in slow distant collisions of two atomic species A and B was already considered in [49] by regarding the interaction between A and B as instantaneous, where in such case, 2CPI proceeds via two-center autoionization and the interaction is transmitted by off-shell photons only. In fact, textbooks on atomic collisions (see, e.g. [81–83]) strongly recommend that such an approach is appropriate for describing slow collisions of light atomic species (in which all the particles involved move with velocities orders of magnitude smaller than the speed of light). However, we have seen that a more complete treatment of 2CPI has to take into account the relativistic retardation effect, which allows for a very efficient (resonant) coupling of the $A - B$ system to the radiation field. This in turn enables the interaction to be transmitted by on-shell photons that dramatically increases its effective range and can profoundly modify the process of 2CPI.

We have compared our calculated rate for 2CPI occurring via the coupling to the radiation field (where the interatomic interaction is transmitted by on-shell photons) to the rate for 2CPI proceeding via two-center autoionization (where the interaction is transmitted by off-shell photons). It was concluded that 2CPI in slow collisions when it involves a (relatively) large energy transfer from atom B to A, can be strongly dominated by the exchange of on-shell photons. Since 2CPI competes with the direct photoionization of A by the laser field, we have further discussed the relative effectiveness of these two processes. Here, we have shown that in case 2CPI is already dominated by the exchange of on-shell photons, it can more than double the rate for ionization of atoms A.

Besides, we have considered the process of 2CPI as a competition between the Coulomb and Breit interactions, where the latter can be transmitted by the exchange of on-shell photons, corresponding to 2CPI when it is driven by the coupling to the radiation field. In this case, the Breit interaction can become very efficient and even dominate over the Coulomb interaction.

Our findings are not exclusive to the process of collisional 2CPI but are more general, because the coupling to the radiation field may strongly affect collisions in which one of the atomic species enters the collision in an excited state, independent of how the excitation occurs (for instance, instead of photoexcitation by a laser field one may have impact excitation by charged particles in a plasma). This immediately follows from the structure of the cross section (3.88) for 2CPI via the coupling to the radiation field. The general part of this cross section is described by the second factor in (3.88) and does not depend on a particular excitation mechanism for atom B.

Another point worth mentioning is that in collisional 2CPI the retardation effect can become tremendously more important than in the process of two-center ion impact ionization of a weakly bound diatomic system, the latter of which was discussed in Chapter 2. This may be explained by comparing the particle distances characteristic of the respective ionization processes. Regarding collisional 2CPI via on-shell photon exchange, by far the main contribution to the ionization cross section comes from extremely far collisions with absolute values of the impact parameter reaching macroscopic sizes $\sim$ 1 cm. In contrast, concerning two-center impact ionization of weakly bound systems, even the largest dimers in question, such as the Li-He system with a mean distance between Li and He of $\approx$ 53 a. u., do not exceed linear spatial extensions of $\lesssim 10^2$ a. u. Now, reminding that retardation strongly affects the interatomic interaction when the distance R between the interacting particles is relatively large, $R \gg c/\omega \sim 10^2$ a. u. (where $\omega \sim 1$ a. u. is a typical electronic transition energy for the diatomic systems under consideration), it is evident that retardation can be crucial for 2CPI in slow atomic collisions but is practically negligible for two-center ion impact ionization of weakly bound diatomic systems.

To conclude this study on the radiation-field-driven ionization in laser-assisted slow atomic collisions, we give a brief outlook on possible experimental verification of our theoretical predictions. For example, the effects predicted in this study can be tested in experiments where a beam of slowly moving projectile atoms or ions (e.g. H, H^-, Mg^+, Ca^+, Ti^+, Fe^+, Sr^+, Ba^+) penetrates (or passes close by) a cold He gas target that is exposed to a weak laser field resonant to the $1\,^1S_0 - 2\,^1P_1$ dipole transition in He. Such experiments may also be performed at considerably larger impact energies (up to several MeV/u), for which we expect that the coupling of the collision system to the radiation field still plays a crucial role in ionization of the projectiles. However, this expectation needs to be refined in future studies, where the various competing ionization channels occurring at larger collision velocities should be compared in detail.

Part II

Formation of Antimatter Ions in Interatomic Attachment Reactions

4 Introduction and Preliminary Remarks

4.1 Historical Background and Motivation

Until today, we do not fully understand the asymmetry between matter and antimatter in the Universe. This makes the study of antimatter in theory and experiment a very important subject from the point of view of fundamental physics.

The first theoretical postulation of antimatter goes back to Dirac who proposed the antiparticle to the electron e^- in 1931 [90]. Shortly afterwards, the positron e^+ was experimentally discovered in 1932 [91, 92]. Just over 20 years later, the antiproton $\bar{p}$ was discovered in 1955 [93, 94]. In the following decades, the amounts of positrons and antiprotons that could be produced in laboratories had grown steadily. Naturally, the next goal was to produce the simplest atom of antimatter, the antihydrogen atom $\bar{H}$, which is an $\bar{p}e^+$ bound system. The first attempts for producing small amounts of $\bar{H}$ were made at CERN [95] and Fermilab [96] in the mid and late 90's. In both experiments, $\bar{H}$ was produced in relativistic collisions of antiprotons with a matter atomic species, involving the creation of e^+e^- pairs and the subsequent capture of e^+ by $\bar{p}$. While antiatoms could be successfully observed for the first time, the main issue in these experiments was that the produced $\bar{H}$ were far too fast to capture them in magnetic traps and perform experiments with them.

However, in 2002, the ATHENA Collaboration at CERN was able to produce substantial amounts of $\bar{H}$ at very low energies by mixing cooled and trapped antiprotons and positrons in a cryogenic environment at temperatures ~ 10 K [97]. At the low temperatures and high positron densities present in the ATHENA experiment, $\bar{H}$ is dominantly formed via the three-body reaction $e^+ + e^+ + \bar{p} \rightarrow \bar{H} + e^+$ (see, e.g. [98, 99]), in which one of the positrons is captured by the antiproton whereas the other positron carries away the energy excess. In 2004, the ATRAP

A. Jacob, *Relativistic Effects in Interatomic Ionization Processes and Formation of Antimatter Ions in Interatomic Attachment Reactions*,
https://doi.org/10.1007/978-3-658-43891-3_4

Collaboration at CERN introduced another experiment based on the capture reaction $Ps^* + \bar{p} \rightarrow \bar{H}^* + e^-$, in which low energy excited antihydrogen is formed via resonant charge exchange between excited positronium Ps (Ps is a bound e^+e^- system) and cold antiprotons [100].

Once the produced antihydrogen atoms reach temperatures $\lesssim 0.5$ K, they can be confined in magnetic traps, which was first achieved by the ALPHA Collaboration [101] at CERN in 2010/11 [102, 103]. This enables one to perform high precision experiments with $\bar{H}$ in a controlled environment (see, e.g. [104–106]).

One of the main interests in experiments involving antimatter is the verification of the CPT symmetry which is a fundamental property of the Standard Model of particle physics. The CPT theorem states that the laws of physics are invariant under the combined discrete operations of charge conjugation (C), parity (P) and time reversal (T). A direct consequence of the CPT theorem is that every particle has an antiparticle with equal mass, spin and total lifetime but opposite charge and magnetic moment (for more details see, e.g. [99]). Therefore, antihydrogen, the simplest and currently only accessible antiatom, is an ideal system for studying the CPT symmetry by comparing atomic interactions in antihydrogen and hydrogen atoms under identical experimental conditions. In 2018, the ALPHA Collaboration measured the $1S - 2S$ transition in $\bar{H}$ and concluded that their results are consistent with the CPT symmetry at a relative precision of 2×10^{-12} [107].

Another key interest in laboratory studies of antimatter is measuring the Earth's local gravitational force exerted on antiparticles as a test of the WEP (weak equivalence principle). In simple terms, the WEP states that in a uniform gravitational field all bodies, independent from their composition, fall with the exact same acceleration (for more details see, e.g. [108]). This implies that the Earth's local gravitational acceleration g is the same for matter and antimatter particles. So far, the gravitational interaction between matter and antimatter systems has not been measured directly. The AEGIS experiment [109] at CERN aims to perform a direct measurement of g on antihydrogen. For this purpose, low energy antihydrogen atoms in excited states are formed via the capture reaction $Ps^* + \bar{p} \rightarrow \bar{H}^* + e^-$. Afterwards, a pulsed cold $\bar{H}^*$ beam is created by Stark acceleration and passes through a Moiré deflectometer in which the free fall of the antihydrogen atoms is measured [110].

The GBAR experiment [111] at CERN also plans to perform free fall measurements with cold $\bar{H}$ in the Earth's gravitational field. Here, they will use the positive ion of antihydrogen $\bar{H}^+$, which is an $\bar{p}e^+e^+$ bound system and the antimatter counterpart of the negative ion of hydrogen H^-, as an intermediate particle. First, a cold antiproton beam penetrates a cloud of (excited) positronium and low energy $\bar{H}^+$ is produced via the two successive charge exchange collisions $Ps + \bar{p} \rightarrow \bar{H} + e^-$ and $Ps + \bar{H} \rightarrow \bar{H}^+ + e^-$. Afterwards, the $\bar{H}^+$ ions are accumulated and sympathetically cooled to $\sim 10\ \mu$K in an ion trap. Then, a laser pulse is applied to the trap region and induces photodetachment of the loosely bound positron in $\bar{\mathrm{H}}^+$. The remaining neutral $\bar{\mathrm{H}}$ will begin to fall down from the ion trap and measurements of the free fall of antihydrogen atoms can be carried out [112].

In this thesis, we consider the formation of the positive ion of antihydrogen $\bar{\mathrm{H}}^+$. From the point of view of theoretical atomic physics there is a general interest in finding new mechanisms for the formation of $\bar{\mathrm{H}}^+$ ions (see, e.g. [113–115]). Moreover, the GBAR experiment, in which $\bar{\mathrm{H}}^+$ will be used as an intermediate particle in free fall experiments on $\bar{\mathrm{H}}$ in the gravitational field of the Earth, shows that theory on efficiently producing $\bar{\mathrm{H}}^+$ ions is of great importance for experimental purposes as well (see, e.g. [116–119]).

4.2 Overview of $\bar{\mathrm{H}}^+$ Formation Mechanisms

There exist two main pathways for producing positive ions of antihydrogen directly from antihydrogen atoms. The first one involves the charge exchange collision $\mathrm{Ps} + \bar{\mathrm{H}} \rightarrow \bar{\mathrm{H}}^+ + e^-$, where a bound e^+ is captured by $\bar{\mathrm{H}}$ to form the $\bar{\mathrm{H}}^+$ ion (see, e.g. [115, 117–119]). The second one is based on free positrons which are incident on $\bar{\mathrm{H}}$ atoms and $\bar{\mathrm{H}}^+$ ions can be produced either via radiative or nonradiative attachment of e^+ to $\bar{\mathrm{H}}$. In the present work, we focus on $\bar{\mathrm{H}}^+$ formation according to the second pathway, including the radiative attachment mechanisms

$$\begin{aligned}
&\text{(i)} \quad e^+ + \bar{\mathrm{H}} \rightarrow \bar{\mathrm{H}}^+ + \hbar\omega_k, \\
&\text{(ii)} \quad e^+ + \bar{\mathrm{H}} + N\hbar\omega_0 \rightarrow \bar{\mathrm{H}}^+ + (N+1)\hbar\omega_0, \\
&\text{(iii)} \quad e^+ + \bar{\mathrm{H}} + B \rightarrow \bar{\mathrm{H}}^+ + B^* \rightarrow \bar{\mathrm{H}}^+ + B + \hbar\omega_k
\end{aligned}$$

as well as the nonradiative three-body reactions

$$\begin{aligned}
&\text{(iv)} \quad e^- + e^+ + \bar{\mathrm{H}} \rightarrow e^- + \bar{\mathrm{H}}^+, \\
&\text{(v)} \quad e^+ + e^+ + \bar{\mathrm{H}} \rightarrow e^+ + \bar{\mathrm{H}}^+.
\end{aligned}$$

The radiative formation mechanisms (i)–(iii) share photoemission as their key signature. Reaction (i) is spontaneous radiative attachment, in which $\bar{\mathrm{H}}^+$ is formed due to spontaneous emission of a photon with frequency ω_k by a positron incident on $\bar{\mathrm{H}}$. When positrons are incident on $\bar{\mathrm{H}}$ atoms in the presence of a laser field with frequency ω_0, the formation of $\bar{\mathrm{H}}^+$ ions can also proceed via induced emission of a photon with frequency ω_0. This process is described by mechanism (ii) and called (laser-)induced radiative attachment. Reaction (iii) is two-center dileptonic attachment, which becomes possible in the presence of a neighboring (matter) atom B and where an incident positron is attached to $\bar{\text{H}}$ via resonant transfer of excess energy – driven by the two-center positron-electron (dileptonic) interaction – to B which, as a result, undergoes a transition into an excited state. Subsequently, B relaxes through spontaneous emission of a photon with frequency ω_k and the two-center system becomes stable implying the formation of the $\bar{\text{H}}^+$ ion [120].

Mechanism (iv) is electron-assisted three-body attachment, in which $\bar{\text{H}}^+$ is formed when free positrons are incident on $\bar{\text{H}}$ embedded in a gas of low energy ($\approx$ meV) electrons and positron capture by $\bar{\text{H}}$ proceeds via the positron-electron interaction where the electron takes the energy excess. If the incident electron in mechanism (iv) is replaced by a second incident positron, one positron is attached to $\bar{\text{H}}$ – due to the positron-positron interaction – while the other positron carries away the energy excess. The latter process is characterized by reaction (v) and termed positron-assisted three-body attachment [121].

The main goal of this study is a comparative consideration of the radiative and nonradiative formation mechanisms (i)–(v) in the range of incident positron energies from sub-meV to eV. It will be shown that for positron energies $\lesssim 0.1$ eV, electron-assisted three-body attachment (iv) can strongly dominate over the radiative mechanisms (i)–(iii). In addition, for positron energies $\simeq 1$ eV, two-center dileptonic attachment (ii) and induced radiative attachment (iii) can be much more efficient than spontaneous radiative attachment (i) and electron-assisted three-body attachment (iv). Moreover, we will see that over the whole range of positron energies under consideration, positron-assisted three-body attachment (v) has vanishingly small formation rates.

Part II of this thesis is essentially organized as follows. In the next Section, we introduce the bound state of the $\bar{\text{H}}^+$ ion that will be used throughout this work. Afterwards, in Chapters 5 & 6, we consider the theoretical framework of the radiative attachment mechanisms (i)–(iii) and the nonradiative three-body reactions (iv)–(v), respectively, and obtain formulas for the corresponding $\bar{\text{H}}^+$ formation rates. Numerical results and a detailed comparative discussion of the attachment mechanisms are given in Chapter 7. Finally, we summarize our main findings in Chapter 8.

Atomic units (see overview on p. xxiii) are used throughout if not stated otherwise.

4.3 The Bound State of $\bar{\mathrm{H}}^+$

Concerning the attachment of a positron to antihydrogen via the mechanisms (i)–(v), we regard the positive ion of antihydrogen as an effectively single-positron system. Here, the positron which is initially bound to the antiproton is treated as passive and forms together with the latter a single-rigid body that produces an external (short-range) field which acts on the incident active positron. Thus, the identical positrons are considered as strongly asymmetrical, being subdivided into active and passive. Note that such an approach has been quite successful in the treatment of electron (positron) detachment from H^- ($\bar{H}^+$) by photoabsorption and of its time-inverse process of spontaneous radiative attachment of an electron (positron) to H ($\bar{H}$) (see, e.g. [113, 114] and references therein) as well as in the treatment of electron detachment from H^- by ion impact [122]. Furthermore, a similar approach was used for describing the single ionization of He atoms by charged particles [123, 124] and by a laser field [125].

Considering $\bar{H}^+$ as an effectively single-positron system, we can approximate its bound state by

$$\phi_b(\boldsymbol{r}_p) = N\frac{e^{-\alpha r_p} - e^{-\beta r_p}}{r_p}, \tag{4.1}$$

where $\boldsymbol{r}_p$ is the coordinate of the positron with respect to the antiproton. Further, $N = \sqrt{\frac{\alpha\beta(\alpha+\beta)}{2\pi(\beta-\alpha)^2}}$, $\alpha = 0.235$ a.u. ($\varepsilon_b = -\alpha^2/2 = -0.0275$ a.u. ≈ -0.748 eV is the binding energy) and $\beta = 0.913$ a.u. The wave function (4.1) was obtained by employing a nonlocal separable potential of Yamaguchi [126] for describing a short-range effective interaction of the active positron with the core of the $\bar{H}^+$ ion.

To get an idea about the accuracy of our theoretical treatments of the $\bar{H}^+$ formation mechanisms (i)–(v) using the wavefunction (4.1), we take as an example the spontaneous radiative attachment of a positron to $\bar{H}$, which will be discussed in Section 5.1. Employing (4.1), our calculations for spontaneous radiative attachment provide a cross section that has basically the same shape but is about 30% larger compared with the results of a more accurate approach [114], in which a 200-term two-positron wave function was applied to describe the bound state of $\bar{H}^+$. Such accuracy is quite sufficient for this study, as we focus on order of magnitude effects regarding the competing attachment mechanisms (i)–(v).

Theory of Radiative Attachment

5

This chapter provides a detailed insight into the theoretical treatments of the radiative attachment mechanisms which are spontaneous and (laser-)induced radiative attachment as well as two-center dileptonic attachment. For each of these processes, we derive the formation rate of positive ions of antihydrogen per unit of time and per antihydrogen atom. The following chapter is mainly based on results published initially in Ref. [120].

5.1 Spontaneous Radiative Attachment

Radiative recombination of an electron with a positive ion via emission of a photon has been studied in detail in the past with energies of the incident electrons ranging from below 1 eV to relativistic values (see, e.g. [127–130] and references therein). Note that radiative recombination is the time-inverse process of photoionization (the latter of which is considered in Chapter 3).

Radiative attachment of an electron to a neutral atom is essentially similar to radiative recombination and is subject to the same fundamental mechanism, which is the interaction of the electron-atom system with the radiation field. The theoretical consideration of spontaneous radiative attachment of an electron to hydrogen (see, e.g. [20, 131, 132]) and of a positron to antihydrogen (see, e.g. [113, 114]) is basically identical from the point of view of Quantum Electrodynamics. Therefore, both processes may be based on the same treatment. Although the spontaneous radiative attachment of a positron to antihydrogen has already been studied, we include the calculation of its reaction rate into this work for completeness and consistency.

A. Jacob, *Relativistic Effects in Interatomic Ionization Processes and Formation of Antimatter Ions in Interatomic Attachment Reactions*,
https://doi.org/10.1007/978-3-658-43891-3_5

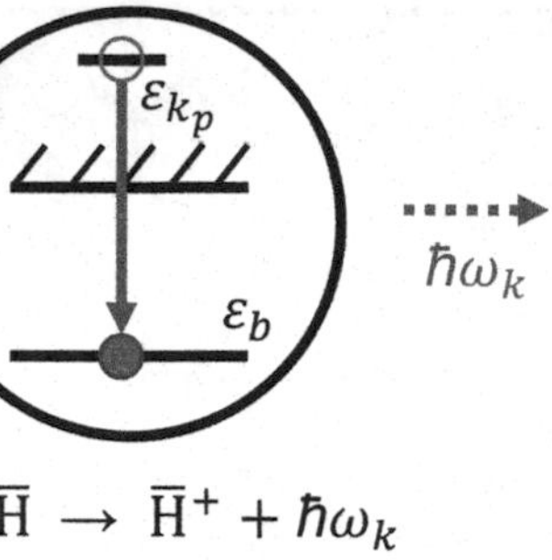

Fig. 5.1 Scheme of spontaneous radiative attachment (SRA). (This figure was originally published in Ref. [120])

We consider spontaneous radiative attachment as an effectively single-positron process. This means that we treat the $\bar{\mathrm{H}}^{+}$ ion as an effectively single-positron system, in which the attached positron is a weakly bound outer positron that moves in the short-range field of the ionic core, which is regarded as a single-rigid body formed by the antiproton and the initially bound positron.

Let us consider an environment where free positrons e^{+} are incident on antihydrogen atoms $\bar{\mathrm{H}}$. In this case, the positive ion of antihydrogen $\bar{\mathrm{H}}^{+}$ can be formed via spontaneous radiative attachment, in which a free e^{+} with kinetic energy ε_{k_p} is captured by $\bar{\mathrm{H}}$ into the ground state of $\bar{\mathrm{H}}^{+}$ with energy ε_b and the energy release is taken away by emission of a photon with energy $\hbar\omega_k$ (see Fig. 5.1 for illustration).

We choose a reference frame in which $\bar{\mathrm{H}}$ is at rest and take the position of its nucleus (the antiproton) as the origin. In this frame, we can describe the attachment process by the Schrödinger equation

$$i\frac{\partial \Psi}{\partial t} = \hat{H}(t)\Psi \tag{5.1}$$

with the total Hamiltonian

$$\hat{H}(t) = \hat{H}_a + \hat{H}_\gamma + \hat{V}_\gamma(t). \tag{5.2}$$

In (5.2),

$$\hat{H}_a = \frac{(\hat{\boldsymbol{p}}_{\boldsymbol{r}_p})^2}{2} - \frac{1}{r_p} \tag{5.3}$$

is the free Hamiltonian of the ($e^+ + \bar{H}$) system, where $\boldsymbol{r}_p$ and $\hat{\boldsymbol{p}}_{\boldsymbol{r}_p}$ are the coordinate and momentum operator of the positron with respect to the antiproton, respectively. In addition,

$$\hat{H}_\gamma = \sum_{\lambda=1,2} \omega_k \hat{a}^\dagger_{\boldsymbol{k}\lambda} \hat{a}_{\boldsymbol{k}\lambda} \tag{5.4}$$

describes the Hamiltonian of the (single-mode) quantized radiation field with the wave vector $\boldsymbol{k}$, the angular frequency $\omega_k = ck$, where c is the speed of light ($c \approx$ 137 a.u.), as well as the creation operator $\hat{a}^\dagger_{\boldsymbol{k}\lambda}$ and annihilation operator $\hat{a}_{\boldsymbol{k}\lambda}$. The two polarization directions of the field are denoted by $\lambda = 1, 2$. Moreover, in (5.2),

$$\hat{V}_\gamma(t) = -\frac{1}{c}\hat{\boldsymbol{A}}(\boldsymbol{r}_p, t) \cdot \hat{\boldsymbol{p}}_{\boldsymbol{r}_p} + \frac{1}{2c^2}\hat{\boldsymbol{A}}^2(\boldsymbol{r}_p, t) \tag{5.5}$$

is the interaction between the ($e^+ + \bar{H}$) system and the radiation field. The latter can be described by the (single-mode) quantized vector potential

$$\hat{\boldsymbol{A}}(\boldsymbol{r}_p, t) = \sqrt{\frac{2\pi c^2}{V_{ph}\omega_k}} \boldsymbol{e}_{\boldsymbol{k}\lambda} \left[\hat{a}_{\boldsymbol{k}\lambda} e^{i(\boldsymbol{k}\cdot\boldsymbol{r}_p - \omega_k t)} + \text{H.c.}\right]. \tag{5.6}$$

Here, $\boldsymbol{e}_{\boldsymbol{k}\lambda}$ ($\lambda = 1, 2$) are the unit polarization vectors ($\boldsymbol{e}_{\boldsymbol{k}1} \cdot \boldsymbol{e}_{\boldsymbol{k}2} = 0$, $\boldsymbol{e}_{\boldsymbol{k}\lambda} \cdot \boldsymbol{k} = 0$) and V_{ph} is the normalization volume for the field. Further, we treat the interaction $\hat{V}_\gamma(t)$ in the dipole approximation, i.e. $\boldsymbol{k} \cdot \boldsymbol{r}_p \approx 0$.
The initial (Ψ_i) and final (Ψ_f) states of the total system, ($e^+ + \bar{H}$) + radiation field, read

$$\begin{aligned}\Psi_i &= \phi_{\boldsymbol{k}_p}(\boldsymbol{r}_p) e^{-i\varepsilon_{k_p} t} \times |0_{\boldsymbol{k}\lambda}\rangle, \\ \Psi_f &= \phi_b(\boldsymbol{r}_p) e^{-i\varepsilon_b t} \times |1_{\boldsymbol{k}\lambda}\rangle,\end{aligned} \tag{5.7}$$

where $\phi_{\boldsymbol{k}_p}$ is the continuum state of the incident positron, which has an asymptotic momentum $\boldsymbol{k}_p$ (as is seen in the rest frame of $\bar{H}$) with corresponding kinetic energy $\varepsilon_{k_p} = \boldsymbol{k}_p^2/2$, and ϕ_b is the bound state of the $\bar{H}^+$ ion. Besides, $|0_{\boldsymbol{k}\lambda}\rangle$ and $|1_{\boldsymbol{k}\lambda}\rangle$ are the states of the radiation field before and after the emission of the photon, respectively.

From the perspective of the incident positron, the (attractive) Coulomb field of the antiproton is largely screened by the presence of the bound (anti-)atomic positron in $\bar{H}$. Therefore, we can describe, in very good approximation, the incident positron by a plane wave

$$\phi_{k_p}(r_p) = \frac{e^{ik_p \cdot r_p}}{\sqrt{V_p}}. \tag{5.8}$$

Here, V_p is the normalization volume for the positron. Further, following the consideration in Section 4.3, we can approximate the bound state ϕ_b of the $\bar{\text{H}}^+$ ion by the wavefunction (4.1).

Using the first order of time-dependent perturbation theory in the interaction between the ($e^+ + \bar{\text{H}}$) system and the radiation field, the transition amplitude for spontaneous radiative attachment can be written as

$$a_{SRA} = -i \int_{-\infty}^{\infty} dt \, \langle \Psi_f | \hat{V}_\gamma(t) | \Psi_i \rangle. \tag{5.9}$$

Next, we insert (5.5) and (5.7) into (5.9), perform the time integration and obtain

$$a_{SRA} = ic\sqrt{\frac{(2\pi)^3}{V_p V_{ph} \omega_k}} \delta(\varepsilon_{k_p} - \varepsilon_b - \omega_k) \langle \phi_b(r_p) | e_{k\lambda} \cdot \hat{p}_{r_p} | e^{ik_p \cdot r_p} \rangle, \tag{5.10}$$

where the delta function reflects the energy conservation, $\omega_k = \varepsilon_{k_p} - \varepsilon_b$, of the attachment process. Taking into account that $\hat{p}_{r_p} | e^{ik_p \cdot r_p} \rangle = k_p | e^{ik_p \cdot r_p} \rangle$, the amplitude (5.10) becomes

$$a_{SRA} = ic\sqrt{\frac{(2\pi)^3}{V_p V_{ph} \omega_k}} \delta(\varepsilon_{k_p} - \varepsilon_b - \omega_k) e_{k\lambda} \cdot k_p \langle \phi_b(r_p) | e^{ik_p \cdot r_p} \rangle. \tag{5.11}$$

The result of the remaining space integral $\langle \phi_b(r_p) | e^{ik_p \cdot r_p} \rangle$ can be easily derived and is given by

$$\begin{aligned} \langle \phi_b(r_p) | e^{ik_p \cdot r_p} \rangle &= \int d^3 r_p \, N \frac{e^{-\alpha r_p} - e^{-\beta r_p}}{r_p} e^{ik_p \cdot r_p} \\ &= 4\pi N \frac{\beta^2 - \alpha^2}{(\alpha^2 + k_p^2)(\beta^2 + k_p^2)}. \end{aligned} \tag{5.12}$$

Inserting (5.12) into (5.11), the transition amplitude for spontaneous radiative attachment reads

$$a_{SRA} = \frac{i\sqrt{2^7 \pi^5} N c (\beta^2 - \alpha^2) \delta(\varepsilon_{k_p} - \varepsilon_b - \omega_k)}{\sqrt{V_p V_{ph} \omega_k} (\alpha^2 + k_p^2)(\beta^2 + k_p^2)} e_{k\lambda} \cdot k_p. \tag{5.13}$$

The formation rate of $\bar{\mathrm{H}}^+$ ions per unit of time (per $\bar{\mathrm{H}}$) for spontaneous radiative attachment is obtained by calculating the quantity

$$\mathcal{R}_{SRA} = \frac{V_{ph}}{(2\pi)^3} \sum_{\lambda=1,2} \int d^3k \, \frac{|a_{SRA}|^2}{T}. \tag{5.14}$$

Here, the time duration T is of the order of the interaction time. We insert (5.13) into (5.14), employ the identity $\left[2\pi\,\delta(\varepsilon_{k_p}-\varepsilon_b-\omega_k)\right]^2 = 2\pi\,\delta(\varepsilon_{k_p}-\varepsilon_b-\omega_k)T$ and choose the unit polarization vectors as $\boldsymbol{e}_{k1} = \boldsymbol{e}_{\vartheta_k} = (\cos\vartheta_k \cos\varphi_k, \cos\vartheta_k \sin\varphi_k, -\sin\vartheta_k)$ and $\boldsymbol{e}_{k2} = \boldsymbol{e}_{\varphi_k} = (-\sin\varphi_k, \cos\varphi_k, 0)$, where $\boldsymbol{e}_{\vartheta_k}$ and $\boldsymbol{e}_{\varphi_k}$ are the polar and azimuthal unit vectors, respectively, when expressing the wave vector $\boldsymbol{k}$ in spherical coordinates. Afterwards, the rate (5.14) can be written as

$$\mathcal{R}_{SRA} = \frac{8\pi N^2 c^2 (\beta^2-\alpha^2)^2 k_p^2}{V_p(\alpha^2+k_p^2)^2(\beta^2+k_p^2)^2} \int d^3k \, \frac{\delta(\varepsilon_{k_p}-\varepsilon_b-\omega_k)}{\omega_k} \sin^2\vartheta_k. \tag{5.15}$$

Now, we carry out the $\boldsymbol{k}$-integral in (5.15) by using $\omega_k = ck$ and $\delta(\varepsilon_{k_p}-\varepsilon_b-\omega_k) = \delta(k - \frac{\varepsilon_{k_p}-\varepsilon_b}{c})/c$ and get

$$\mathcal{R}_{SRA} = \frac{64\pi^2 N^2 (\beta^2-\alpha^2)^2 k_p^2 (\varepsilon_{k_p}-\varepsilon_b)}{3V_p c(\alpha^2+k_p^2)^2(\beta^2+k_p^2)^2}. \tag{5.16}$$

Expressing the normalization volume V_p for the incident positron in terms of the corresponding number density n_p of positrons according to $V_p = 1/n_p$, the final result for the formation rate of $\bar{\mathrm{H}}^+$ ions per unit of time (per $\bar{\mathrm{H}}$) for spontaneous radiative attachment is given by

$$\mathcal{R}_{SRA} = \frac{64\pi^2 n_p N^2 (\beta^2-\alpha^2)^2 k_p^2 (\varepsilon_{k_p}-\varepsilon_b)}{3c(\alpha^2+k_p^2)^2(\beta^2+k_p^2)^2}. \tag{5.17}$$

5.2 (Laser-)induced Radiative Attachment

In a situation, in which free positrons e^+ are incident on antihydrogen atoms $\bar{\mathrm{H}}$ in the presence of a laser field having the frequency ω_0, the formation of the positive ion of antihydrogen $\bar{\mathrm{H}}^+$ can also take place via induced emission of a photon with energy $\hbar\omega_0$ (see Fig. 5.2 for illustration). We refer to this process as (laser-)induced radiative attachment.

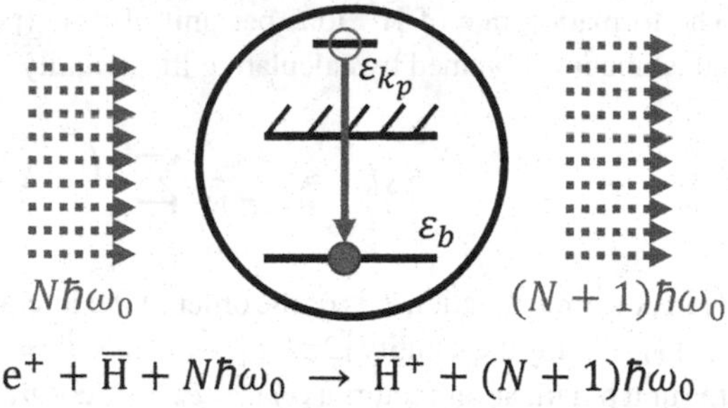

Fig. 5.2 Scheme of (laser-)induced radiative attachment (LIRA). (This figure was originally published in Ref. [120])

The laser field is considered as a classical monochromatic electromagnetic wave of linear polarization in the dipole approximation, $\boldsymbol{F}(t) = \boldsymbol{F}_0 \sin(\omega_0 t)$ with $|\boldsymbol{F}_0| = F_0$ the strength of the field. It shall efficiently stimulate attachment without destroying the produced $\bar{H}^+$ ions (and obviously without destroying the $\bar{H}$ atoms themselves). In order to accomplish this, we assume that the laser field is sufficiently weak, $F_0 \ll F_a$ with F_a the typical atomic field which is produced by the ionic core and acts on the loosely bound outer positron in $\bar{H}^+$. Moreover, it is also assumed that the laser frequency ω_0 is resonantly tuned to the positron transitions which lead to the formation of the $\bar{H}^+$ ion, i.e. $\omega_0 \approx \varepsilon_{k_p} - \varepsilon_b$.

For describing the process of (laser-)induced radiative attachment (as well as of (laser-)induced resonant scattering), we can use the Schrödinger equation

$$i\frac{\partial \Psi}{\partial t} = (\hat{H}_a + \hat{W}(t))\Psi \tag{5.18}$$

with $\hat{H}_a$ the Hamiltonian of the field-free $(e^+ + \bar{H})$ system, given by (5.3), and

$$\hat{W}(t) = -\frac{1}{c}\boldsymbol{A}_L(t)\cdot\hat{\boldsymbol{p}}_{\boldsymbol{r}_p} \tag{5.19}$$

the interaction between the incident positron and the laser field. In (5.19),

$$\boldsymbol{A}_L(t) = \frac{c\boldsymbol{F}_0}{\omega_0}\cos(\omega_0 t) \tag{5.20}$$

is the (classical) vector potential that is associated with the field $\boldsymbol{F}(t)$ when applying the so-called velocity gauge, in which the electric field $\boldsymbol{F}$ is determined solely by the vector potential $\boldsymbol{A}_L$ according to $\boldsymbol{F}(t) = -\frac{1}{c}\frac{\partial \boldsymbol{A}_L(t)}{\partial t}$.

We expand the wave function $|\Psi\rangle$ into the complete set of states

$$|\Psi\rangle = a_b(t)|\phi_b(\boldsymbol{r}_p)\rangle e^{-i\varepsilon_b t} + \int d^3\boldsymbol{k}_p\, b_{\boldsymbol{k}_p}(t)|\phi_{\boldsymbol{k}_p}(\boldsymbol{r}_p)\rangle e^{-i\varepsilon_{\boldsymbol{k}_p} t} \tag{5.21}$$

and insert (5.21) into (5.18). The resulting equation is projected on $\langle\phi_b|$ and $\langle\phi_{\boldsymbol{k}'_p}|$, respectively. Afterwards, we employ the rotating wave approximation (see, e.g. [85]), in which the rapidly oscillating time-dependent terms are dropped. Since the electromagnetic field is assumed to be sufficiently weak, it is further possible to neglect the laser-induced transitions between the continuum states. In addition, we assume that the laser field is switched on suddenly at $t = t_i = 0$. Taking all this into account, the set of equations for the unknown coefficients $a_b(t)$ and $b_{\boldsymbol{k}_p}(t)$ in (5.21) is obtained to be

$$\begin{aligned} i\dot{a}_b(t) &= \int d^3\boldsymbol{k}_p b_{\boldsymbol{k}_p}(t)\mathcal{W}_{b\boldsymbol{k}_p} e^{-i(\varepsilon_{kp}-\varepsilon_b-\omega_0)t}, \\ i\dot{b}_{\boldsymbol{k}_p}(t) &= a_b(t)\mathcal{W}^*_{b\boldsymbol{k}_p} e^{i(\varepsilon_{kp}-\varepsilon_b-\omega_0)t} \end{aligned} \tag{5.22}$$

with the initial conditions $a_b(t = t_i = 0) = 0$ and $b_{\boldsymbol{k}_p}(t = t_i = 0) = \delta^3(\boldsymbol{k}_p - \boldsymbol{k}_{p,0})$, where $\boldsymbol{k}_{p,0}$ is the incident positron momentum. The transition matrix element $\mathcal{W}_{b\boldsymbol{k}_p}$ in (5.22) reads

$$\mathcal{W}_{b\boldsymbol{k}_p} = -\frac{1}{2\omega_0}\langle\phi_b(\boldsymbol{r}_p)|\boldsymbol{F}_0 \cdot \hat{\boldsymbol{p}}_{\boldsymbol{r}_p}|\phi_{\boldsymbol{k}_p}(\boldsymbol{r}_p)\rangle \tag{5.23}$$

and $\mathcal{W}^*_{b\boldsymbol{k}_p}$ is the complex conjugate of $\mathcal{W}_{b\boldsymbol{k}_p}$.

Next, we define $\tilde{b}_{\boldsymbol{k}_p}(t) = b_{\boldsymbol{k}_p}(t)e^{-i(\varepsilon_{kp}-\varepsilon_b-\omega_0)t}$ with $\tilde{b}_{\boldsymbol{k}_p}(t = t_i = 0) = b_{\boldsymbol{k}_p}(t = 0) = \delta^3(\boldsymbol{k}_p - \boldsymbol{k}_{p,0})$. Then, (5.22) becomes

$$\begin{aligned} i\dot{a}_b(t) &= \int d^3\boldsymbol{k}_p\tilde{b}_{\boldsymbol{k}_p}(t)\mathcal{W}_{b\boldsymbol{k}_p}, \\ i\dot{\tilde{b}}_{\boldsymbol{k}_p}(t) - (\varepsilon_{k_p} - \varepsilon_b - \omega_0)\tilde{b}_{\boldsymbol{k}_p}(t) &= a_b(t)\mathcal{W}^*_{b\boldsymbol{k}_p}. \end{aligned} \tag{5.24}$$

The system of equations (5.24) can be solved by using the Laplace transform (see, e.g. [133])

$$L_f(s) = \int_0^\infty dt\ f(t)e^{-st}. \tag{5.25}$$

For this purpose, we first multiply both equations in (5.24) by the factor e^{-st} and subsequently integrate them over the time from $t = 0$ to $t = \infty$ while taking advantage of the initial conditions $a_b(t = 0) = 0$ and $\tilde{b}_{\mathbf{k}_p}(t = 0) = \delta^3(\mathbf{k}_p - \mathbf{k}_{p,0})$, which yields

$$
\begin{aligned}
isL_{a_b}(s) &= \int d^3\mathbf{k}_p L_{\tilde{b}_{\mathbf{k}_p}}(s)\mathcal{W}_{b\mathbf{k}_p}, \\
(is - \varepsilon_{\mathbf{k}_p} + \varepsilon_b + \omega_0)L_{\tilde{b}_{\mathbf{k}_p}}(s) - i\delta^3(\mathbf{k}_p - \mathbf{k}_{p,0}) &= L_{a_b}(s)\mathcal{W}^*_{b\mathbf{k}_p}.
\end{aligned} \tag{5.26}
$$

Solving the set of equations (5.26) for $L_{a_b}(s)$ provides

$$
L_{a_b}(s) = \frac{i\mathcal{W}_{b\mathbf{k}_{p,0}}}{\left(is - \varepsilon_{\mathbf{k}_{p,0}} + \varepsilon_b + \omega_0\right)\left(is - \int d^3\mathbf{k}_p \frac{|\mathcal{W}_{b\mathbf{k}_p}|^2}{is - \varepsilon_{\mathbf{k}_p} + \varepsilon_b + \omega_0}\right)}. \tag{5.27}
$$

The inverse Laplace transform is defined by (see, e.g. [133])

$$
f(t) = \frac{1}{2\pi i}\int_{\gamma - i\infty}^{\gamma + i\infty} ds\, L_f(s)e^{st}, \tag{5.28}
$$

where $\gamma \in \mathbb{R}$ is a constant that exceeds the real part of any of the singular points of $L_f(s)$. Applying (5.28) to (5.27) leads to

$$
a_b(t) = \frac{\mathcal{W}_{b\mathbf{k}_{p,0}}}{2\pi}\int_{\gamma - i\infty}^{\gamma + i\infty} ds \frac{e^{st}}{\left(is - \varepsilon_{\mathbf{k}_{p,0}} + \varepsilon_b + \omega_0\right)\left(is - \int d^3\mathbf{k}_p \frac{|\mathcal{W}_{b\mathbf{k}_p}|^2}{is - \varepsilon_{\mathbf{k}_p} + \varepsilon_b + \omega_0}\right)}. \tag{5.29}
$$

Further, in (5.29), we substitute $z = is \in \mathbb{C}$ and get

$$
a_b(t) = \frac{i\mathcal{W}_{b\mathbf{k}_{p,0}}}{2\pi}\int_{-\infty + i\gamma}^{\infty + i\gamma} dz \frac{e^{-izt}}{\left(z - \varepsilon_{\mathbf{k}_{p,0}} + \varepsilon_b + \omega_0\right)\left(z - \int d^3\mathbf{k}_p \frac{|\mathcal{W}_{b\mathbf{k}_p}|^2}{z - \varepsilon_{\mathbf{k}_p} + \varepsilon_b + \omega_0}\right)}. \tag{5.30}
$$

The integral in (5.30) can be calculated by employing the so-called pole approximation [120], in which

$$\int d^3\boldsymbol{k}_p \frac{|\mathcal{W}_{b\boldsymbol{k}_p}|^2}{z - \varepsilon_{\boldsymbol{k}_p} + \varepsilon_b + \omega_0} \approx \Delta - i\frac{\Gamma}{2}. \tag{5.31}$$

Here,

$$\Delta = \text{P.V.} \int d^3\boldsymbol{k}_p \frac{|\mathcal{W}_{b\boldsymbol{k}_p}|^2}{\varepsilon_b + \omega_0 - \varepsilon_{\boldsymbol{k}_p}} \tag{5.32}$$

is a small energy shift and

$$\Gamma = 2\pi \int d^2\Omega_{\boldsymbol{k}_p} |\mathcal{W}_{b\boldsymbol{k}_p}|^2_{k_p=|\boldsymbol{k}_p|=\sqrt{2(\varepsilon_b+\omega_0)}}, \tag{5.33}$$

where the integration runs over the solid angle $\Omega_{\boldsymbol{k}_p}$ of the positron, is the width of the bound state ϕ_b of the $\bar{\text{H}}^+$ ion due to the interaction with the laser field. The pole approximation is very accurate as long as $\varepsilon_b + \omega_0 \gg \max\{\Delta, \Gamma\}$, which in case of a relatively weak field is well satisfied. Insertion of (5.31) into (5.30) yields

$$a_b(t) = \frac{i\mathcal{W}_{b\boldsymbol{k}_{p,0}}}{2\pi} \int_{-\infty+i\gamma}^{\infty+i\gamma} dz \frac{e^{-izt}}{\left(z - \varepsilon_{\boldsymbol{k}_{p,0}} + \varepsilon_b + \omega_0\right)\left(z - \Delta + i\frac{\Gamma}{2}\right)}. \tag{5.34}$$

Now, we can easily perform the integration in (5.34) by using the Residue theorem and obtain

$$a_b(t) = \frac{\mathcal{W}_{b\boldsymbol{k}_{p,0}} e^{-i(\varepsilon_{\boldsymbol{k}_{p,0}} - \varepsilon_b - \omega_0)t} \left(1 - e^{i(\varepsilon_{\boldsymbol{k}_{p,0}} - \varepsilon_b - \omega_0 + i\frac{\Gamma}{2})t}\right)}{\varepsilon_{\boldsymbol{k}_{p,0}} - \varepsilon_b - \omega_0 + i\frac{\Gamma}{2}}, \tag{5.35}$$

where it is assumed that the small shift Δ is already included into the energy ε_b.

The probability to find the incident positron in the bound state of $\bar{\text{H}}^+$ is given by

$$\begin{aligned} P_{a_b}(t) &= |a_b(t)|^2 \\ &= \frac{|\mathcal{W}_{b\boldsymbol{k}_{p,0}}|^2}{(\varepsilon_{\boldsymbol{k}_{p,0}} - \varepsilon_b - \omega_0)^2 + \frac{\Gamma^2}{4}} \left(1 - 2\cos((\varepsilon_{\boldsymbol{k}_{p,0}} - \varepsilon_b - \omega_0)t)e^{-\frac{\Gamma}{2}t} + e^{-\Gamma t}\right). \end{aligned} \tag{5.36}$$

Note that the factor $\left[(\varepsilon_{\boldsymbol{k}_{p,0}} - \varepsilon_b - \omega_0)^2 + \Gamma^2/4\right]^{-1}$ in (5.36) describes a well-known resonance structure (see, e.g. [39, 40, 44]) having a maximum at the exact resonance $\varepsilon_{\boldsymbol{k}_{p,res}} = \varepsilon_b + \omega_0$. In contrast to spontaneous radiative attachment, (laser-)induced radiative attachment is a resonant process that is only efficient in a very narrow interval of incident positron energies centered at $\varepsilon_{k_{p,res}} = \varepsilon_b + \omega_0$ and having an effective width of a few Γ's.

If the incident positrons do not have a fixed momentum $\boldsymbol{k}_{p,0}$, one may average the probability (5.36) over their momentum distribution function $f(\boldsymbol{k}_{p,0})$ according to

$$\langle P_{a_b}(t)\rangle = \int d^3\boldsymbol{k}_{p,0}\; f(\boldsymbol{k}_{p,0}) P_{a_b}(t). \tag{5.37}$$

In order to derive a simple analytical result for the averaged probability, we make the following assumptions: (i) the polarization of the laser field is chosen along the z-axis ($\boldsymbol{F}_0 = F_0\boldsymbol{e}_z$), (ii) we assume that the incident positrons move along the field polarization ($\boldsymbol{k}_{p,0} = k_{p,0}\boldsymbol{e}_z$), and (iii) their energies are supposed to be uniformly distributed over an interval $I_\varepsilon = [\varepsilon_{k_{p,res}} - \Delta\varepsilon_{k_p}/2, \varepsilon_{k_{p,res}} + \Delta\varepsilon_{k_p}/2]$ that is centered at the resonance energy $\varepsilon_{k_{p,res}} = \varepsilon_b + \omega_0$ and has a width $\Delta\varepsilon_{k_p}$ which is much narrower than the effective energy width ΔE of the $\bar{\mathrm{H}}^+$ continuum ($\Delta E \simeq 1$ eV) but at the same time is much broader than the width Γ of the $\bar{\mathrm{H}}^+$ bound state. Note that in a relatively weak laser field Γ amounts to just a tiny fraction of 1 eV, such that our assumptions on $\Delta\varepsilon_{k_p}$ in (iii) are very well compatible with each other. Taking all this into account, the resulting momentum distribution function $f(\boldsymbol{k}_{p,0})$ for the incident positrons can be written as

$$f(\boldsymbol{k}_{p,0}) = \frac{\delta(1-\cos\vartheta_{\boldsymbol{k}_{p,0}})\chi_{\{\varepsilon_{k_{p,0}}\in I_\varepsilon\}}(\varepsilon_{k_{p,0}})}{2\pi k_{p,0}\Delta\varepsilon_{k_p}}, \tag{5.38}$$

where $\chi_{\{\varepsilon_{k_{p,0}}\in I_\varepsilon\}}(\varepsilon_{k_{p,0}})$ is the indicator function which takes the value 1 for $\varepsilon_{k_{p,0}} \in I_\varepsilon$ and 0 otherwise. Inserting (5.38) as well as (5.36) into (5.37) and rewriting the integral over $k_{p,0} = |\boldsymbol{k}_{p,0}|$ into an integral over the positron energy $\varepsilon_{k_{p,0}} = k_{p,0}^2/2$, we arrive at

$$\begin{aligned}\langle P_{a_b}(t)\rangle = {} & \frac{1}{2\pi\Delta\varepsilon_{k_p}}\int_{I_\varepsilon} d\varepsilon_{k_{p,0}}\, \frac{\left(1-2\cos((\varepsilon_{k_{p,0}}-\varepsilon_b-\omega_0)t)e^{-\frac{\Gamma}{2}t}+e^{-\Gamma t}\right)}{(\varepsilon_{k_{p,0}}-\varepsilon_b-\omega_0)^2+\frac{\Gamma^2}{4}} \\ & \times \int d^2\Omega_{\boldsymbol{k}_{p,0}}\,\delta(1-\cos\vartheta_{\boldsymbol{k}_{p,0}})|\mathcal{W}_{b\boldsymbol{k}_{p,0}}|^2. \end{aligned} \tag{5.39}$$

First, we focus on the solid angle integral in (5.39). When performing the integration over the polar angle $\vartheta_{\boldsymbol{k}_{p,0}}$, the resulting integrand $|\mathcal{W}_{b\boldsymbol{k}_{p,0}}|^2_{\vartheta_{\boldsymbol{k}_{p,0}}=0} =$

$|\mathcal{W}_{b\boldsymbol{k}_{p,0}}|^2_{\boldsymbol{k}_{p,0}=k_{p,0}\boldsymbol{e}_z}$ does not depend on the azimuthal angle $\varphi_{\boldsymbol{k}_{p,0}}$. Thus, the subsequent integration over $\varphi_{\boldsymbol{k}_{p,0}}$ simply yields $2\pi|\mathcal{W}_{b\boldsymbol{k}_{p,0}}|^2_{\boldsymbol{k}_{p,0}=k_{p,0}\boldsymbol{e}_z}$ and the averaged probability $\langle P_{a_b}(t)\rangle$ becomes

$$\langle P_{a_b}(t)\rangle = \frac{1}{\Delta\varepsilon_{k_p}} \int_{I_\varepsilon} d\varepsilon_{k_{p,0}} \frac{\left(1 - 2\cos((\varepsilon_{k_{p,0}} - \varepsilon_b - \omega_0)t)e^{-\frac{\Gamma}{2}t} + e^{-\Gamma t}\right)}{(\varepsilon_{k_{p,0}} - \varepsilon_b - \omega_0)^2 + \frac{\Gamma^2}{4}} \times |\mathcal{W}_{b\boldsymbol{k}_{p,0}}|^2_{\boldsymbol{k}_{p,0}=k_{p,0}\boldsymbol{e}_z}. \tag{5.40}$$

In (5.40), the resonant factor $g(\varepsilon_{k_{p,0}}) = \left[(\varepsilon_{k_{p,0}} - \varepsilon_b - \omega_0)^2 + \Gamma^2/4\right]^{-1}$ only contributes to the integral in a very narrow interval of energies of the incident positron centered at the maximum $\varepsilon_{k_{p,res}} = \varepsilon_b + \omega_0$ of the function $g(\varepsilon_{k_{p,0}})$ and having a width of a few Γ's (which is of course much smaller than the width $\Delta\varepsilon_{k_p}$ of the interval I_ε). Within this interval, $\varepsilon_{k_{p,res}} - \Gamma \lesssim \varepsilon_{k_{p,0}} \lesssim \varepsilon_{k_{p,res}} + \Gamma$, the function $g(\varepsilon_{k_{p,0}})$ varies much faster in $\varepsilon_{k_{p,0}}$ than the quantity $|\mathcal{W}_{b\boldsymbol{k}_{p,0}}|^2_{\boldsymbol{k}_{p,0}=k_{p,0}\boldsymbol{e}_z}$ and we can treat, in an approximate manner, the latter as a constant evaluated at the resonance energy $\varepsilon_{k_{p,res}} = \varepsilon_b + \omega_0$. Furthermore, since the contribution of the integrand is negligibly small outside the small interval $\varepsilon_{k_{p,res}} - \Gamma \lesssim \varepsilon_{k_{p,0}} \lesssim \varepsilon_{k_{p,res}} + \Gamma$, we can extend the integration boundaries to 0 to ∞. Then, (5.40) reads

$$\langle P_{a_b}(t)\rangle = \frac{|\mathcal{W}_{b\boldsymbol{k}_{p,0}}|^2_{\boldsymbol{k}_{p,0}=k_{p,res}\boldsymbol{e}_z}}{\Delta\varepsilon_{k_p}} \times \int_0^\infty d\varepsilon_{k_{p,0}} \frac{\left(1 - 2\cos((\varepsilon_{k_{p,0}} - \varepsilon_b - \omega_0)t)e^{-\frac{\Gamma}{2}t} + e^{-\Gamma t}\right)}{(\varepsilon_{k_{p,0}} - \varepsilon_b - \omega_0)^2 + \frac{\Gamma^2}{4}} \tag{5.41}$$

with $k_{p,res} = \sqrt{2(\varepsilon_b + \omega_0)}$. The remaining integral in (5.41) can be solved quite easily by substituting $u = \varepsilon_{k_{p,0}} - \varepsilon_b - \omega_0$ and afterwards taking advantage of the fact that the resulting integrand only contributes to the integral in a very narrow interval $-\Gamma \lesssim u \lesssim \Gamma$, so that the lower integration boundary $-(\varepsilon_b + \omega_0)$ can be extended to $-\infty$. Then, the averaged probability $\langle P_{a_b}(t)\rangle$ to find the incident positron in the bound state of $\bar{\mathrm{H}}^+$ is obtained to be

$$\langle P_{a_b}(t)\rangle = \frac{|\mathcal{W}_{b\boldsymbol{k}_{p,0}}|^2_{\boldsymbol{k}_{p,0}=k_{p,res}\boldsymbol{e}_z}}{\Delta\varepsilon_{k_p}} \frac{2\pi}{\Gamma}\left(1 - e^{-\Gamma t}\right). \tag{5.42}$$

The time derivative of (5.42) yields the (averaged) formation rate of $\bar{\text{H}}^+$ ions per unit of time (per $\bar{\text{H}}$) that is

$$\langle \mathcal{R}_{LIRA} \rangle = \frac{d}{dt}\langle P_{a_b}(t) \rangle = \frac{|\mathcal{W}_{b\boldsymbol{k}_{p,0}}|^2_{\boldsymbol{k}_{p,0}=k_{p,res}\boldsymbol{e}_z}}{\Delta\varepsilon_{k_p}} 2\pi e^{-\Gamma t}. \tag{5.43}$$

The rate (5.43) contains the transition matrix element $\mathcal{W}_{b\boldsymbol{k}_{p,0}}$ from (5.23). For its evaluation, we use the same continuum and bound states which have already been used in the consideration of spontaneous radiative attachment in Section 5.1 and which are given by (5.8) and (4.1), respectively. Recalling that $\boldsymbol{F}_0 = F_0\boldsymbol{e}_z$, the quantity $|\mathcal{W}_{b\boldsymbol{k}_{p,0}}|^2_{\boldsymbol{k}_{p,0}=k_{p,res}\boldsymbol{e}_z}$ can be written as

$$\begin{aligned} |\mathcal{W}_{b\boldsymbol{k}_{p,0}}|^2_{\boldsymbol{k}_{p,0}=k_{p,res}\boldsymbol{e}_z} &= \left| -\frac{1}{2\omega_0}\langle \phi_b(\boldsymbol{r}_p)|\boldsymbol{F}_0 \cdot \hat{\boldsymbol{p}}_{\boldsymbol{r}_p}|\phi_{\boldsymbol{k}_{p,0}}(\boldsymbol{r}_p)\rangle \right|^2_{\boldsymbol{k}_{p,0}=k_{p,res}\boldsymbol{e}_z} \\ &= \frac{F_0^2}{4\omega_0^2 V_p}|\langle \phi_b(\boldsymbol{r}_p)|\boldsymbol{e}_z \cdot \hat{\boldsymbol{p}}_{\boldsymbol{r}_p}|e^{ik_{p,res}\boldsymbol{e}_z\cdot\boldsymbol{r}_p}\rangle|^2. \end{aligned} \tag{5.44}$$

Next, we apply $\hat{\boldsymbol{p}}_{\boldsymbol{r}_p}|e^{ik_{p,res}\boldsymbol{e}_z\cdot\boldsymbol{r}_p}\rangle = k_{p,res}\boldsymbol{e}_z|e^{ik_{p,res}\boldsymbol{e}_z\cdot\boldsymbol{r}_p}\rangle$ and (5.44) becomes

$$|\mathcal{W}_{b\boldsymbol{k}_{p,0}}|^2_{\boldsymbol{k}_{p,0}=k_{p,res}\boldsymbol{e}_z} = \frac{F_0^2 k_{p,res}^2}{4\omega_0^2 V_p}|\langle \phi_b(\boldsymbol{r}_p)|e^{ik_{p,res}\boldsymbol{e}_z\cdot\boldsymbol{r}_p}\rangle|^2. \tag{5.45}$$

The result of $\langle \phi_b(\boldsymbol{r}_p)|e^{ik_{p,res}\boldsymbol{e}_z\cdot\boldsymbol{r}_p}\rangle$ in (5.45) can be obtained from (5.12) and we arrive at

$$|\mathcal{W}_{b\boldsymbol{k}_{p,0}}|^2_{\boldsymbol{k}_{p,0}=k_{p,res}\boldsymbol{e}_z} = \frac{4\pi^2 N^2 F_0^2 k_{p,res}^2 (\beta^2-\alpha^2)^2}{\omega_0^2 V_p (\alpha^2 + k_{p,res}^2)^2 (\beta^2 + k_{p,res}^2)^2}. \tag{5.46}$$

Inserting (5.46) into (5.43) and expressing the normalization volume V_p of the incident positron via the corresponding number density n_p of positrons as $V_p = 1/n_p$, the (averaged) $\bar{\text{H}}^+$ formation rate per unit of time (per $\bar{\text{H}}$) for (laser-)induced radiative attachment reads

$$\langle \mathcal{R}_{LIRA} \rangle = \frac{8\pi^3 n_p N^2 F_0^2 k_{p,res}^2 (\beta^2-\alpha^2)^2}{\Delta\varepsilon_{k_p}\omega_0^2 (\alpha^2 + k_{p,res}^2)^2 (\beta^2 + k_{p,res}^2)^2} e^{-\Gamma t}. \tag{5.47}$$

As can be concluded from (5.47), the formation rate substantially diminishes for $\Gamma t \simeq 1$ and already essentially vanishes for $\Gamma t \gg 1$. The reason for this is that in the presence of a laser field, besides induced radiative attachment, photodetachment (of the loosely bound positron in $\bar{\mathrm{H}}^{+}$) also occurs. Hence, with increasing the population of the bound state of $\bar{\mathrm{H}}^{+}$, the attachment and detachment events may balance each other after some time, resulting in a zero net formation rate. Let the laser pulse have a time duration T, where $0 \leq t \leq T$. In order to efficiently produce $\bar{\mathrm{H}}^{+}$ ions via (laser-)induced radiative attachment during the whole pulse, the duration T should be sufficiently small so that $\Gamma T \ll 1$ and consequently $\Gamma t \ll 1$.

Note that the formation of $\bar{\mathrm{H}}^{+}$ ions via induced radiative attachment involving a weak laser field has also been considered independently in [134]. However, compared with our theoretical approach to this process, in which the Schrödinger equation describing the induced attachment is solved by applying the rotating wave approximation, the approach in [134] is quite different. It treats the attachment by taking advantage of the first order of time-dependent perturbation theory in the interaction between the incident positron and the weak laser field.

5.3 Two-Center Dileptonic Attachment

Let us now consider attachment of a positron e^{+} to an antihydrogen atom $\bar{\mathrm{H}}$ proceeding in the presence of a neighboring (matter) atom B. For the moment, we ignore possible annihilation channels and also other processes that might occur in an environment in which antimatter is embedded in matter. These will be discussed in detail in Section 7.2.

At first, we assume a fixed interatomic distance R_0 ($R_0 \gg 1$ a.u.) between $\bar{\mathrm{H}}$ and B. If the energy excess $\omega_{\bar{\mathrm{H}}^{+}} = \varepsilon_{k_p} - \varepsilon_b$ in the process of $e^{+} + \bar{\mathrm{H}}$ attachment is close to an excitation energy $\omega_B = \epsilon_e - \epsilon_g$ of a dipole-allowed transition in atom B (ϵ_e and ϵ_g are the energies of the excited and ground state of B, respectively), the released energy can be transferred – via the two-center positron-electron (dileptonic) interaction – to atom B, which, as a result, undergoes a transition into an excited state. Afterwards, B radiatively decays to its initial (ground) state by emission of a photon with energy $\hbar\omega_k$ and the two-center system, $(e^{+} + \bar{\mathrm{H}}) + B$, becomes stable, meaning that the $\bar{\mathrm{H}}^{+}$ ion has been formed. A scheme of this process, termed two-center dileptonic attachment, is shown in Fig. 5.3.

In the 'matter' case of two-center dielectronic recombination (attachment), in which an electron e^{-} recombines with (is attached to) a positive charged ion (a neutral atom) A in the presence of a neighbor atom B driven by the two-center electron-electron interaction, it is well-known [39, 40, 45] that, because of its reso-

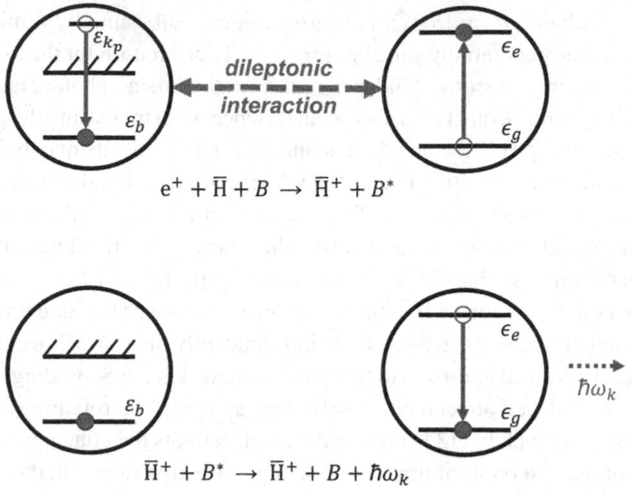

Fig. 5.3 Scheme of two-center dileptonic attachment (2CDA). (This figure was originally published in Ref. [120])

nant nature, the two-center channel can enhance the corresponding recombination (attachment) rate by orders of magnitude as compared with the case where B is absent and recombination (attachment) is only possible via direct photoemission from the $(e^- + A)$ system. We mention that two-center dielectronic recombination (attachment) is the time-inverse process of two-center photoionization (photodetachment), which was considered for colliding atomic species A and B in Chapter 3.

Suppose now that free positrons and a beam of $\bar{H}$ move in a (dilute) gas of atoms B. As previously mentioned, two-center dileptonic attachment relies on an energy transfer which is resonant to a dipole transition in B. However, the relative motion of $\bar{H}$ and B results in uncertainty in positron and electron transition energies (as they are seen by the corresponding collision partner), effectively broadening them. For this reason, the efficiency of the two-center attachment channel is limited to low velocity collisions, in which the velocity v of $\bar{H}$ with regard to B is much less than 1 a.u. [41, 135].

In a recent study [41, 135], we have considered two-center dielectronic recombination (attachment) when electrons and a beam of slow positive ions (neutral atoms) A move in a gas of atomic species B. In the same manner as for the direct correspondence between spontaneous radiative attachment of an electron to an atom and of a positron to an antiatom (see Section 5.1), the results obtained in [41, 135] can be straightforwardly adapted to the process of two-center dileptonic attachment of a positron to $\bar{\mathrm{H}}$ proceeding in an environment where free positrons and a slow $\bar{\mathrm{H}}$ beam move in a gas of atoms B.

Using the results from [41, 135], the formation rate of $\bar{\mathrm{H}}^+$ ions per unit of time (per $\bar{\mathrm{H}}$) for two-center dileptonic attachment is given by

$$\mathcal{R}_{2CDA} = \mathcal{R}_{SRA} \times \frac{9\pi}{4} \frac{n_B}{v\, b_{min}^2} \frac{c^6 \Gamma_r^B}{\omega_{\bar{\mathrm{H}}^+}^3 \omega_B^3} \eta^2 \Big\{ \sin^2 \vartheta_{k_p} K_1^2(\eta) + (1 + \cos^2 \vartheta_{k_p}) \eta K_0(\eta) K_1(\eta) \Big\}. \quad (5.48)$$

Here, $\mathcal{R}_{SRA}$ is the rate for spontaneous radiative attachment of a positron to $\bar{\mathrm{H}}$ from (5.17), n_B is the density of atoms B, b_{min} is the minimum impact parameter in the $\bar{\mathrm{H}} - B$ collisions, Γ_r^B is the radiative width of the excited state of atom B, $\eta = |\omega_{\bar{\mathrm{H}}^+} - \omega_B| b_{min}/v$ and ϑ_{k_p} is the incident positron angle (which is counted from the collision velocity $\boldsymbol{v}$). Further, $K_j(x)$ $(j = 0, 1)$ are the modified Bessel functions [87].

Note that within the theoretical approach of two-center dielectronic recombination (attachment) in [41, 135], only contributions to the two-center channel from distant collisions, in which $b_{min} \gg 1$ a.u., were taken into account. Therefore, the rate (5.48) represents just a lower boundary of the $\bar{\mathrm{H}}^+$ formation rate for two-center dileptonic attachment.

The functions $K_j(x)$ $(j = 0, 1)$ diverge at $x \to 0$ and decrease exponentially at $x > 1$ [87]. Thus, in distant ($b_{min} \gg 1$ a.u.) and low velocity ($v \ll 1$ a.u.) collisions, the formation of $\bar{\mathrm{H}}^+$ ions via two-center attachment is most favorable, according to (5.48), when the energy of the incident positrons lies within the small interval $\varepsilon_b + \omega_B - v/b_{min} \lesssim \varepsilon_{k_p} \lesssim \varepsilon_b + \omega_B + v/b_{min}$ centered at $\varepsilon_{k_{p,res}} = \varepsilon_b + \omega_B$ and having the width $\delta\varepsilon_{k_p} \sim v/b_{min}$. Because the quantity v/b_{min} is typically orders of magnitude larger than the radiative width Γ_r^B, we can conclude that the relative motion of $\bar{\mathrm{H}}$ and B strongly smears out the 'static' resonance condition $\varepsilon_b + \omega_B - \Gamma_r^B \lesssim \varepsilon_{k_p} \lesssim \varepsilon_b + \omega_B + \Gamma_r^B$ [41, 135], in case where the distance between $\bar{\mathrm{H}}$ and B is fixed, and leads to a much broader range of 'quasiresonance' energies of the incident positron.

Theory of Nonradiative Attachment 6

This chapter provides a detailed insight into the theoretical treatments of the nonradiative three-body reactions, which are electron-assisted three-body attachment and positron-assisted three-body attachment. For each of these processes, we derive the formation rate of positive ions of antihydrogen per unit of time and per antihydrogen atom. The following chapter is mainly based on results published initially in Ref. [121].

6.1 Electron-assisted Three-body Attachment

In an environment, in which free positrons e^+ and electrons e^- move in a close vicinity of antihydrogen atoms $\bar{\mathrm{H}}$, attachment of a positron to $\bar{\mathrm{H}}$ may occur due to the positron-electron interaction, where the energy excess is taken away by an electron. We call this process electron-assisted three-body attachment and its schematic representation is pictured in Fig. 6.1, where ε_{k_e} and $\varepsilon_{k'_e}$ are the energies of the incident and outgoing electron, respectively.

For the moment, we ignore the possibility of positron-electron annihilation, whose influence on the three-body reaction will be discussed in detail in Section 7.2.

The considered attachment mechanism is a four-body problem that is rather difficult to handle from a theoretical point of view. Hence, we may treat this process as an effectively three-body problem by regarding the initially bound positron as passive, forming together with the antiproton a single-rigid body which produces an external (short-range) field that acts on the active positron and the electron. Such

Supplementary Information The online version contains supplementary material available at https://doi.org/10.1007/978-3-658-43891-3_6.

A. Jacob, *Relativistic Effects in Interatomic Ionization Processes and Formation of Antimatter Ions in Interatomic Attachment Reactions*,
https://doi.org/10.1007/978-3-658-43891-3_6

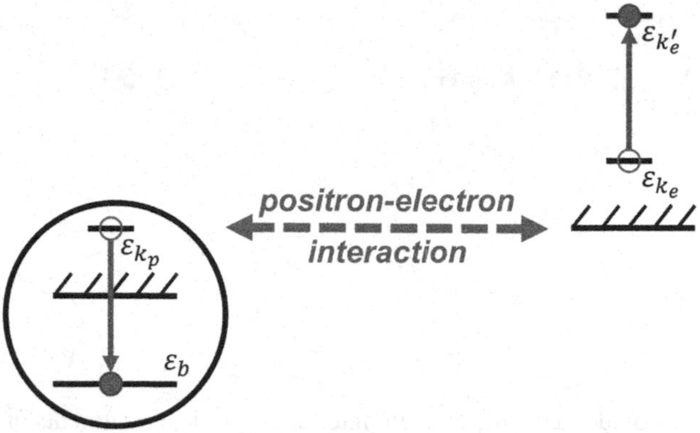

$$e^- + e^+ + \bar{H} \rightarrow e^- + \bar{H}^+$$

Fig. 6.1 Scheme of electron-assisted three-body attachment (3BAe). (This figure was originally published in Refs. [120, 121])

an approximation, in which identical particles are considered as sharply asymmetrically, subdividing them into active and passive, was already used for the treatments of spontaneous and (laser-)induced radiative attachment in Section 5.1 and Section 5.2, where we pointed out that this approximation has been fairly successful in describing several atomic processes involving two-electron (positron) systems.

Let us choose a reference frame in which $\bar{H}$ is at rest and take the position of the antiproton as the origin. We start our consideration of the three-body attachment with the exact transition amplitude in the post form which can be written as (see, e.g. [129])

$$a_{3BAe} = -i \int_{-\infty}^{\infty} dt \left\langle \left(\hat{H} - i \frac{\partial}{\partial t} \right) \chi_f(t) \middle| \Psi_i^{(+)}(t) \right\rangle, \tag{6.1}$$

where $\Psi_i^{(+)}(t)$ is the exact solution of the full Schrödinger equation (whose Hamiltonian $\hat{H}$ includes all the interactions) that satisfies the incoming boundary condition and $\chi_f(t)$ is the final asymptotic state that is a solution of the Schrödinger equation with the corresponding asymptotic Hamiltonian $\hat{H}_f$.

The Hamiltonian $\hat{H}$ in (6.1) reads

$$\hat{H} = \frac{(\hat{\boldsymbol{p}}_{r_e})^2}{2} + \hat{V}_e(\boldsymbol{r}_e) + \frac{(\hat{\boldsymbol{p}}_{r_p})^2}{2} + \hat{V}_p(\boldsymbol{r}_p) + \hat{V}_{ep}(\boldsymbol{r}_e, \boldsymbol{r}_p) \tag{6.2}$$

with $\boldsymbol{r}_e$ and $\hat{\boldsymbol{p}}_{r_e}$ the coordinate and momentum operator for the incident electron with respect to the antiproton, respectively. Further, $\hat{V}_e(\boldsymbol{r}_e)$ is the interaction between the electron and H̄. Similarly, $\hat{V}_p(\boldsymbol{r}_p)$ describes the interaction between the positron and H̄. Besides, $\hat{V}_{ep}(\boldsymbol{r}_e, \boldsymbol{r}_p) = -\frac{1}{|\boldsymbol{r}_e - \boldsymbol{r}_p|}$ is the Coulomb interaction between the incident electron and positron.

Taking advantage of the fact that the Coulomb interaction between the incident electron and positron is much stronger than the interactions of these particles with the (neutral) antihydrogen, we can approximate the exact state $\Psi_i^{(+)}(t)$ according to

$$\Psi_i^{(+)}(\boldsymbol{r}_e, \boldsymbol{r}_p, t) = e^{i\boldsymbol{P}\cdot\boldsymbol{R}} \psi_{\boldsymbol{\kappa}}^{(+)}(\boldsymbol{r}) e^{-i(\varepsilon_{k_p} + \varepsilon_{k_e})t}. \tag{6.3}$$

In (6.3), $\boldsymbol{R} = (\boldsymbol{r}_e + \boldsymbol{r}_p)/2$ is the coordinate of the center-of-mass of the incident electron-positron pair and $\boldsymbol{P} = \boldsymbol{k}_e + \boldsymbol{k}_p$ is their total momentum, where $\boldsymbol{k}_e$ is the asymptotic momentum of the incident electron (as is seen in the rest frame of H̄) with corresponding kinetic energy $\varepsilon_{k_e} = \boldsymbol{k}_e^2/2$. In addition,

$$\psi_{\boldsymbol{\kappa}}^{(+)}(\boldsymbol{r}) = \frac{1}{\sqrt{V_p V_e}} e^{\frac{\pi}{2\kappa}} \Gamma\left(1 - \frac{i}{\kappa}\right) e^{i\boldsymbol{\kappa}\cdot\boldsymbol{r}} F\left(\frac{i}{\kappa}, 1, i(\kappa r - \boldsymbol{\kappa}\cdot\boldsymbol{r})\right) \tag{6.4}$$

is the Coulomb wave function which describes the relative motion of the incident electron and positron. Here, $\boldsymbol{r} = \boldsymbol{r}_e - \boldsymbol{r}_p$ is the relative coordinate of the electron-positron pair, $\boldsymbol{\kappa} = (\boldsymbol{k}_e - \boldsymbol{k}_p)/2$, V_e is the electron normalization volume and $F(a, b, z)$ is the confluent hypergeometric function [87]. Consequently, the state (6.3) describes the motion of the incident electron and positron by fully accounting for the long-range Coulomb interaction between them while neglecting their interaction with the (neutral) antihydrogen.

We choose the final state $\chi_f(t)$ as

$$\chi_f(\boldsymbol{r}_e, \boldsymbol{r}_p, t) = \phi_b(\boldsymbol{r}_p) \varphi_f(\boldsymbol{r}_e, \boldsymbol{r}_p) e^{-i(\varepsilon_b + \varepsilon_{k_e'})t} \tag{6.5}$$

with $\phi_b(\boldsymbol{r}_p)$ the (undistorted) bound state of the positron which was attached to H̄ and $\varphi_f(\boldsymbol{r}_e, \boldsymbol{r}_p)$ the state of the outgoing electron that moves in the field of the $\bar{\text{H}}^+$

ion formed. We treat $\bar{\mathrm{H}}^+$ as an effectively one-positron system with the positron attached being a weakly bound outer positron moving in the short-range field of the (single-body) ionic core. Then, following the consideration in Section 4.3, we can approximate the state ϕ_b by the wavefunction (4.1). Furthermore, we suppose that the state φ_f is described by the Coulomb wave function which takes into account the influence of the Coulomb interaction between the outgoing electron and the bound positron on each of these two particles and which is given by

$$\varphi_{\boldsymbol{k}'_e}^{(-)}(\boldsymbol{r}) = \frac{1}{\sqrt{V_e}} e^{\frac{\pi}{2k'_e}} \Gamma\left(1 + \frac{i}{k'_e}\right) e^{i\boldsymbol{k}'_e \cdot \boldsymbol{r}} F\left(-\frac{i}{k'_e}, 1, -i(k'_e r + \boldsymbol{k}'_e \cdot \boldsymbol{r})\right), \quad (6.6)$$

where $\boldsymbol{k}'_e$ is the asymptotic momentum of the outgoing electron with the corresponding kinetic energy $\varepsilon_{k'_e} = \boldsymbol{k}_e'^2/2$.

Using the Hamiltonian (6.2), in which the weak interaction $\hat{V}_e(\boldsymbol{r}_e)$ between the electron and the (neutral) antihydrogen is being neglected, and using (6.5), the quantity $\left(\hat{H} - i\frac{\partial}{\partial t}\right)\chi_f(t)$ in the amplitude (6.1) becomes

$$\begin{aligned}\left(\hat{H} - i\frac{\partial}{\partial t}\right)\chi_f(t) = &-i\frac{\partial}{\partial t}\phi_b(\boldsymbol{r}_p)\varphi_{\boldsymbol{k}'_e}^{(-)}(\boldsymbol{r})e^{-i(\varepsilon_b+\varepsilon_{k'_e})t} \\ &+\left[\frac{(\hat{\boldsymbol{p}}_{\boldsymbol{r}_e})^2}{2} - \frac{1}{r}\right]\phi_b(\boldsymbol{r}_p)\varphi_{\boldsymbol{k}'_e}^{(-)}(\boldsymbol{r})e^{-i(\varepsilon_b+\varepsilon_{k'_e})t} \\ &+\left[\frac{(\hat{\boldsymbol{p}}_{\boldsymbol{r}_p})^2}{2} + \hat{V}_p(\boldsymbol{r}_p)\right]\phi_b(\boldsymbol{r}_p)\varphi_{\boldsymbol{k}'_e}^{(-)}(\boldsymbol{r})e^{-i(\varepsilon_b+\varepsilon_{k'_e})t}. \quad (6.7)\end{aligned}$$

By taking advantage of the eigenvalue equations $\left[\frac{(\hat{\boldsymbol{p}}_{\boldsymbol{r}_p})^2}{2} + \hat{V}_p(\boldsymbol{r}_p)\right]\phi_b(\boldsymbol{r}_p) = \varepsilon_b\phi_b(\boldsymbol{r}_p)$ and $\left[\frac{(\hat{\boldsymbol{p}}_{\boldsymbol{r}_e})^2}{2} - \frac{1}{r}\right]\varphi_{\boldsymbol{k}'_e}^{(-)}(\boldsymbol{r}) = \left[\frac{(\hat{\boldsymbol{p}}_{\boldsymbol{r}})^2}{2} - \frac{1}{r}\right]\varphi_{\boldsymbol{k}'_e}^{(-)}(\boldsymbol{r}) = \varepsilon_{k'_e}\varphi_{\boldsymbol{k}'_e}^{(-)}(\boldsymbol{r})$, it can be easily shown that (6.7) yields

$$\begin{aligned}\left(\hat{H} - i\frac{\partial}{\partial t}\right)\chi_f(t) = \Bigg\{&\left[\varepsilon_{k'_e} + \frac{1}{r}\right]\phi_b(\boldsymbol{r}_p)\varphi_{\boldsymbol{k}'_e}^{(-)}(\boldsymbol{r}) \\ &-\left[\hat{\boldsymbol{p}}_{\boldsymbol{r}_p}\phi_b(\boldsymbol{r}_p)\right]\cdot\left[\hat{\boldsymbol{p}}_{\boldsymbol{r}_e}\varphi_{\boldsymbol{k}'_e}^{(-)}(\boldsymbol{r})\right]\Bigg\}e^{-i(\varepsilon_b+\varepsilon_{k'_e})t}. \quad (6.8)\end{aligned}$$

Now, we insert (6.3) and (6.8) into the transition amplitude (6.1), perform the integration over the time t and obtain

$$a_{3BAe} = -2\pi i\delta(\varepsilon_{k_p} - \varepsilon_b - (\varepsilon_{k'_e} - \varepsilon_{k_e}))\left(\mathcal{M}_1 + \varepsilon_{k'_e}\mathcal{M}_2 - \mathcal{M}_3\right). \quad (6.9)$$

Here,

$$\mathcal{M}_1 = \left\langle \frac{1}{r}\phi_b(\boldsymbol{r}_p)\varphi^{(-)}_{\boldsymbol{k}'_e}(\boldsymbol{r}) \middle| e^{i\boldsymbol{P}\cdot\boldsymbol{R}}\psi^{(+)}_{\boldsymbol{\kappa}}(\boldsymbol{r})\right\rangle,$$

$$\mathcal{M}_2 = \left\langle \phi_b(\boldsymbol{r}_p)\varphi^{(-)}_{\boldsymbol{k}'_e}(\boldsymbol{r}) \middle| e^{i\boldsymbol{P}\cdot\boldsymbol{R}}\psi^{(+)}_{\boldsymbol{\kappa}}(\boldsymbol{r})\right\rangle,$$

$$\mathcal{M}_3 = \left\langle \left[\hat{\boldsymbol{p}}_{\boldsymbol{r}_p}\phi_b(\boldsymbol{r}_p)\right]\cdot\left[\hat{\boldsymbol{p}}_{\boldsymbol{r}_e}\varphi^{(-)}_{\boldsymbol{k}'_e}(\boldsymbol{r})\right] \middle| e^{i\boldsymbol{P}\cdot\boldsymbol{R}}\psi^{(+)}_{\boldsymbol{\kappa}}(\boldsymbol{r})\right\rangle. \tag{6.10}$$

In (6.10), we first consider the space integral $\mathcal{M}_1$ which reads

$$\mathcal{M}_1 = \int d^3\boldsymbol{r}_p \int d^3\boldsymbol{r}_e \left(\frac{1}{r}\phi_b(\boldsymbol{r}_p)\varphi^{(-)}_{\boldsymbol{k}'_e}(\boldsymbol{r})\right)^* e^{i\boldsymbol{P}\cdot\boldsymbol{R}}\psi^{(+)}_{\boldsymbol{\kappa}}(\boldsymbol{r}). \tag{6.11}$$

Rewriting the integral over the electron coordinate $\boldsymbol{r}_e$ in (6.11) into an integral over the relative coordinate $\boldsymbol{r} = \boldsymbol{r}_e - \boldsymbol{r}_p$ of the electron-positron pair provides

$$\mathcal{M}_1 = \left(\int d^3\boldsymbol{r}_p\, e^{i\boldsymbol{P}\cdot\boldsymbol{r}_p}\phi_b^*(\boldsymbol{r}_p)\right)\left(\int d^3\boldsymbol{r}\, \frac{e^{i\boldsymbol{P}\cdot\frac{\boldsymbol{r}}{2}}}{r}\psi^{(+)}_{\boldsymbol{\kappa}}(\boldsymbol{r})[\varphi^{(-)}_{\boldsymbol{k}'_e}(\boldsymbol{r})]^*\right) = \mathcal{I}_1\,\mathcal{I}_2. \tag{6.12}$$

The calculation of the integral $\mathcal{I}_1$ in (6.12) is quite simple and yields

$$\mathcal{I}_1 = \int d^3\boldsymbol{r}_p\, e^{i\boldsymbol{P}\cdot\boldsymbol{r}_p} N \frac{e^{-\alpha r_p} - e^{-\beta r_p}}{r_p} = 4\pi N \frac{\beta^2-\alpha^2}{(\alpha^2+P^2)(\beta^2+P^2)}. \tag{6.13}$$

The integral $\mathcal{I}_2$ in (6.12) requires a more careful treatment. Applying the Coulomb wave functions (6.4) and (6.6) and introducing the infinitesimal small positive parameter λ ($\lambda \to 0^+$), $\mathcal{I}_2$ can be written as

$$\mathcal{I}_2 = \frac{1}{\sqrt{V_p V_e}} e^{\frac{\pi}{2\kappa}}\Gamma\left(1-\frac{i}{\kappa}\right)\frac{1}{\sqrt{V_e}} e^{\frac{\pi}{2k'_e}}\Gamma\left(1-\frac{i}{k'_e}\right)\mathcal{J} \tag{6.14}$$

with

$$\mathcal{J} = \int d^3\boldsymbol{r}\, e^{-\lambda r}\frac{e^{i\boldsymbol{q}\cdot\boldsymbol{r}}}{r} F\left(\frac{i}{\kappa}, 1, i(\kappa r - \boldsymbol{\kappa}\cdot\boldsymbol{r})\right) F\left(\frac{i}{k'_e}, 1, i(k'_e r + \boldsymbol{k}'_e\cdot\boldsymbol{r})\right), \tag{6.15}$$

where $\boldsymbol{q} = \frac{\boldsymbol{P}}{2} + \boldsymbol{\kappa} - \boldsymbol{k}'_e = \boldsymbol{k}_e - \boldsymbol{k}'_e$. An integral of the same form as $\mathcal{J}$ also appears in the theory of Bremsstrahlung and it was calculated in [136]. Then, using the results of [136], the integral $\mathcal{J}$ in (6.15) is obtained to be

$$\mathcal{J} = \frac{2\pi}{\tilde{\alpha}} e^{-\frac{\pi}{\kappa}} \left(\frac{\tilde{\alpha}}{\tilde{\gamma}}\right)^{\frac{i}{\kappa}} \left(\frac{\tilde{\gamma}+\tilde{\delta}}{\tilde{\gamma}}\right)^{-\frac{i}{k'_e}} F\left(1-\frac{i}{\kappa}, \frac{i}{k'_e}, 1, z\right). \tag{6.16}$$

Here, $z = \frac{\tilde{\alpha}\tilde{\delta}-\tilde{\beta}\tilde{\gamma}}{\tilde{\alpha}(\tilde{\gamma}+\tilde{\delta})}$, $\tilde{\alpha} = (q^2+\lambda^2)/2$, $\tilde{\beta} = \boldsymbol{k}'_e \cdot \boldsymbol{q} - i\lambda k'_e$, $\tilde{\gamma} = \boldsymbol{\kappa}\cdot\boldsymbol{q} + i\lambda\kappa - \tilde{\alpha}$, $\tilde{\delta} = \kappa k'_e + \boldsymbol{\kappa}\cdot\boldsymbol{k}'_e - \tilde{\beta}$ and $F(a, b, c, z)$ is the hypergeometric function [87].

Inserting (6.16) into (6.14) and afterwards inserting the resulting expression for $\mathcal{I}_2$ as well as the result for $\mathcal{I}_1$ from (6.13) into (6.12), the quantity $\mathcal{M}_1$ becomes

$$\begin{aligned}\mathcal{M}_1 = 8\pi^2 N \frac{\beta^2-\alpha^2}{(\alpha^2+P^2)(\beta^2+P^2)} \frac{1}{\sqrt{V_p V_e}} e^{-\frac{\pi}{2\kappa}} \Gamma\left(1-\frac{i}{\kappa}\right) \frac{1}{\sqrt{V_e}} e^{\frac{\pi}{2k'_e}} \Gamma\left(1-\frac{i}{k'_e}\right) \\ \times \frac{1}{\tilde{\alpha}} \left(\frac{\tilde{\alpha}}{\tilde{\gamma}}\right)^{\frac{i}{\kappa}} \left(\frac{\tilde{\gamma}+\tilde{\delta}}{\tilde{\gamma}}\right)^{-\frac{i}{k'_e}} F\left(1-\frac{i}{\kappa}, \frac{i}{k'_e}, 1, z\right).\end{aligned} \tag{6.17}$$

Next, we evaluate the space integral $\mathcal{M}_2$ in (6.10) that reads

$$\mathcal{M}_2 = \int d^3\boldsymbol{r}_p \int d^3\boldsymbol{r}_e \left(\phi_b(\boldsymbol{r}_p)\varphi^{(-)}_{\boldsymbol{k}'_e}(\boldsymbol{r})\right)^* e^{i\boldsymbol{P}\cdot\boldsymbol{R}} \psi^{(+)}_{\boldsymbol{\kappa}}(\boldsymbol{r}). \tag{6.18}$$

The integral over the coordinate $\boldsymbol{r}_e$ in (6.18) is rewritten into an integral over the relative coordinate $\boldsymbol{r} = \boldsymbol{r}_e - \boldsymbol{r}_p$ and we arrive at

$$\mathcal{M}_2 = \left(\int d^3\boldsymbol{r}_p\, e^{i\boldsymbol{P}\cdot\boldsymbol{r}_p} \phi_b^*(\boldsymbol{r}_p)\right)\left(\int d^3\boldsymbol{r}\, e^{i\boldsymbol{P}\cdot\frac{\boldsymbol{r}}{2}} \psi^{(+)}_{\boldsymbol{\kappa}}(\boldsymbol{r})[\varphi^{(-)}_{\boldsymbol{k}'_e}(\boldsymbol{r})]^*\right) = \mathcal{I}_1\mathcal{I}_3, \tag{6.19}$$

where the result of the integral $\mathcal{I}_1$ is given by (6.13).

Employing the states (6.4) and (6.6) and, as before, introducing the infinitesimal small positive parameter λ ($\lambda \to 0^+$), the integral $\mathcal{I}_3$ in (6.19) yields

$$\begin{aligned}\mathcal{I}_3 = \frac{1}{\sqrt{V_p V_e}} e^{\frac{\pi}{2\kappa}} \Gamma\left(1-\frac{i}{\kappa}\right) \frac{1}{\sqrt{V_e}} e^{\frac{\pi}{2k'_e}} \Gamma\left(1-\frac{i}{k'_e}\right) \\ \times \int d^3\boldsymbol{r}\, e^{-\lambda r} e^{i\boldsymbol{q}\cdot\boldsymbol{r}} F\left(\frac{i}{\kappa}, 1, i(\kappa r - \boldsymbol{\kappa}\cdot\boldsymbol{r})\right) F\left(\frac{i}{k'_e}, 1, i(k'_e r + \boldsymbol{k}'_e\cdot\boldsymbol{r})\right).\end{aligned} \tag{6.20}$$

In (6.20), we can recover the integral $\mathcal{J}$ from (6.15) by differentiation with respect to λ. In particular, taking into account that $e^{-\lambda r} = -\frac{\partial}{\partial\lambda}\left(\frac{e^{-\lambda r}}{r}\right)$, (6.20) can be written as

$$\mathcal{I}_3 = -\frac{1}{\sqrt{V_p V_e}} e^{\frac{\pi}{2\kappa}} \Gamma\left(1 - \frac{i}{\kappa}\right) \frac{1}{\sqrt{V_e}} e^{\frac{\pi}{2k_e'}} \Gamma\left(1 - \frac{i}{k_e'}\right)\left(\frac{\partial}{\partial\lambda}\mathcal{J}\right). \tag{6.21}$$

The solution of the integral $\mathcal{J}$ is given by (6.16). Applying the differentiation formula (see, e.g. [87]) $\frac{d}{dz}F(a, b, c, z) = \frac{ab}{c}F(a+1, b+1, c+1, z)$ for the hypergeometric function, the derivative of (6.16) with respect to λ is straightforward and provides

$$\begin{aligned}\frac{\partial}{\partial\lambda}\mathcal{J} = \frac{2\pi}{\tilde{\alpha}} e^{-\frac{\pi}{\kappa}} \left(\frac{\tilde{\alpha}}{\tilde{\gamma}}\right)^{\frac{i}{\kappa}} \left(\frac{\tilde{\gamma}+\tilde{\delta}}{\tilde{\gamma}}\right)^{-\frac{i}{k_e'}} &\left[A_1 F\left(1 - \frac{i}{\kappa}, \frac{i}{k_e'}, 1, z\right)\right. \\ &\left.+A_2 F\left(2 - \frac{i}{\kappa}, \frac{i}{k_e'} + 1, 2, z\right)\right].\end{aligned} \tag{6.22}$$

Here, the quite cumbersome quantities $A_{1,2} = A_{1,2}(\boldsymbol{k}_p, \boldsymbol{k}_e, \boldsymbol{k}_e')$ are shown in Appendix 9.5 in the Electronic Supplementary Material.

Inserting (6.22) into (6.21) and subsequently inserting the resulting expression for $\mathcal{I}_3$ as well as the result for $\mathcal{I}_1$ from (6.13) into (6.19), the quantity $\mathcal{M}_2$ is obtained to be

$$\begin{aligned}\mathcal{M}_2 = &-8\pi^2 N \frac{\beta^2 - \alpha^2}{(\alpha^2 + P^2)(\beta^2 + P^2)} \frac{1}{\sqrt{V_p V_e}} e^{-\frac{\pi}{2\kappa}} \Gamma\left(1 - \frac{i}{\kappa}\right) \frac{1}{\sqrt{V_e}} e^{\frac{\pi}{2k_e'}} \Gamma\left(1 - \frac{i}{k_e'}\right) \\ &\times \frac{1}{\tilde{\alpha}}\left(\frac{\tilde{\alpha}}{\tilde{\gamma}}\right)^{\frac{i}{\kappa}} \left(\frac{\tilde{\gamma}+\tilde{\delta}}{\tilde{\gamma}}\right)^{-\frac{i}{k_e'}} \left[A_1 F\left(1 - \frac{i}{\kappa}, \frac{i}{k_e'}, 1, z\right)\right. \\ &\left.+A_2 F\left(2 - \frac{i}{\kappa}, \frac{i}{k_e'} + 1, 2, z\right)\right].\end{aligned} \tag{6.23}$$

The last (and most complicated) quantity to be determined in (6.10) is the space integral $\mathcal{M}_3$ which reads

$$\mathcal{M}_3 = \int d^3 r_p \int d^3 r_e \left(\left[\hat{\boldsymbol{p}}_{r_p} \phi_b(\boldsymbol{r}_p)\right] \cdot \left[\hat{\boldsymbol{p}}_{r_e} \varphi_{k_e'}^{(-)}(\boldsymbol{r})\right]\right)^* e^{i\boldsymbol{P}\cdot\boldsymbol{R}} \psi_{\kappa}^{(+)}(\boldsymbol{r}). \tag{6.24}$$

Rewriting the integral over the coordinate $\boldsymbol{r}_e$ in (6.24) into an integral over the relative coordinate $\boldsymbol{r} = \boldsymbol{r}_e - \boldsymbol{r}_p$ leads to

$$\mathcal{M}_3 = \left(\int d^3\boldsymbol{r}_p\, e^{i\boldsymbol{P}\cdot\boldsymbol{r}_p}\big[\hat{\boldsymbol{p}}_{\boldsymbol{r}_p}\phi_b(\boldsymbol{r}_p)\big]^*\right)\left(\int d^3\boldsymbol{r}\, e^{i\boldsymbol{P}\cdot\frac{\boldsymbol{r}}{2}}\psi_{\boldsymbol{\kappa}}^{(+)}(\boldsymbol{r})\big[\hat{\boldsymbol{p}}_{\boldsymbol{r}}\varphi_{\boldsymbol{k}'_e}^{(-)}(\boldsymbol{r})\big]^*\right)$$
$$= \mathcal{I}_4\,\mathcal{I}_5. \tag{6.25}$$

Using integration by parts, where $\phi_b(\boldsymbol{r}_p) \to 0$ as $|\boldsymbol{r}_p| \to \infty$, the integral $\mathcal{I}_4$ in (6.25) becomes

$$\mathcal{I}_4 = \boldsymbol{P}\int d^3\boldsymbol{r}_p\, e^{i\boldsymbol{P}\cdot\boldsymbol{r}_p}\phi_b^*(\boldsymbol{r}_p) = \boldsymbol{P}\,\mathcal{I}_1 = \boldsymbol{P}\,4\pi N\frac{\beta^2-\alpha^2}{(\alpha^2+P^2)(\beta^2+P^2)} \tag{6.26}$$

with the result of the integral $\mathcal{I}_1$ given by (6.13).

Now, we turn to the integral $\mathcal{I}_5$ in (6.25), whose evaluation is quite laborious. Applying the states (6.4) and (6.6) and, once again, introducing the infinitesimal small positive parameter λ ($\lambda \to 0^+$), $\mathcal{I}_5$ yields

$$\mathcal{I}_5 = i\frac{1}{\sqrt{V_pV_e}}e^{\frac{\pi}{2\kappa}}\Gamma\left(1-\frac{i}{\kappa}\right)\frac{1}{\sqrt{V_e}}e^{\frac{\pi}{2k'_e}}\Gamma\left(1-\frac{i}{k'_e}\right)$$
$$\times\int d^3\boldsymbol{r}\, e^{-\lambda r}e^{i\left(\frac{\boldsymbol{P}}{2}+\boldsymbol{\kappa}\right)\cdot\boldsymbol{r}}F\left(\frac{i}{\kappa},1,i(\kappa r-\boldsymbol{\kappa}\cdot\boldsymbol{r})\right)$$
$$\times\nabla_{\boldsymbol{r}}\left[e^{-i\boldsymbol{k}'_e\cdot\boldsymbol{r}}F\left(\frac{i}{k'_e},1,i(k'_e r+\boldsymbol{k}'_e\cdot\boldsymbol{r})\right)\right]. \tag{6.27}$$

Employing $\nabla_{\boldsymbol{r}}F\big(i\nu,1,i(k'_e r+\boldsymbol{k}'_e\cdot\boldsymbol{r})\big) = \frac{k'_e}{r}\nabla_{\boldsymbol{k}'_e}\Big[F\big(i\nu,1,i(k'_e r+\boldsymbol{k}'_e\cdot\boldsymbol{r})\big)\Big]_{\nu=\text{const.}}$, where $\nu = 1/k'_e$ is regarded as a constant with respect to $\nabla_{\boldsymbol{k}'_e}$, it is easy to show that the quantity $\nabla_{\boldsymbol{r}}\big[e^{-i\boldsymbol{k}'_e\cdot\boldsymbol{r}}F\big(\frac{i}{k'_e},1,i(k'_e r+\boldsymbol{k}'_e\cdot\boldsymbol{r})\big)\big]$ in (6.27) can be expressed as

$$\nabla_{\boldsymbol{r}}\left[e^{-i\boldsymbol{k}'_e\cdot\boldsymbol{r}}F\left(i\nu,1,i(k'_e r+\boldsymbol{k}'_e\cdot\boldsymbol{r})\right)\right] =$$
$$\left(\frac{k'_e}{r}\nabla_{\boldsymbol{k}'_e}+\frac{k'_e}{r}i\boldsymbol{r}-i\boldsymbol{k}'_e\right)\left[e^{-i\boldsymbol{k}'_e\cdot\boldsymbol{r}}F\left(i\nu,1,i(k'_e r+\boldsymbol{k}'_e\cdot\boldsymbol{r})\right)\right]. \tag{6.28}$$

Inserting (6.28) into (6.27) provides

$$\mathcal{I}_5 = i\frac{1}{\sqrt{V_p V_e}} e^{\frac{\pi}{2\kappa}} \Gamma\left(1 - \frac{i}{\kappa}\right) \frac{1}{\sqrt{V_e}} e^{\frac{\pi}{2k'_e}} \Gamma\left(1 - \frac{i}{k'_e}\right) \Big[k'_e (\nabla_{k'_e} \mathcal{J})$$
$$+ k'_e \int d^3\boldsymbol{r}\, e^{-\lambda r} i\boldsymbol{r} \frac{e^{i\boldsymbol{q}\cdot\boldsymbol{r}}}{r} F\left(\frac{i}{\kappa}, 1, i(\kappa r - \boldsymbol{\kappa}\cdot\boldsymbol{r})\right) F\left(i\nu, 1, i(k'_e r + \boldsymbol{k}'_e \cdot \boldsymbol{r})\right)$$
$$- i\boldsymbol{k}'_e \int d^3\boldsymbol{r}\, e^{-\lambda r} e^{i\boldsymbol{q}\cdot\boldsymbol{r}} F\left(\frac{i}{\kappa}, 1, i(\kappa r - \boldsymbol{\kappa}\cdot\boldsymbol{r})\right) F\left(i\nu, 1, i(k'_e r + \boldsymbol{k}'_e \cdot \boldsymbol{r})\right) \Big], \tag{6.29}$$

where the integral $\mathcal{J}$ in the first term is defined by (6.15). The second and third term of (6.29) can also be expressed in terms of $\mathcal{J}$ by differentiation with respect to $\boldsymbol{q}$ and λ, respectively. Thus, using $i\boldsymbol{r}\frac{e^{i\boldsymbol{q}\cdot\boldsymbol{r}}}{r} = \nabla_{\boldsymbol{q}}(\frac{e^{i\boldsymbol{q}\cdot\boldsymbol{r}}}{r})$ in the second term and $e^{-\lambda r} = -\frac{\partial}{\partial\lambda}(\frac{e^{-\lambda r}}{r})$ in the third term, (6.29) can be written as

$$\mathcal{I}_5 = i\frac{1}{\sqrt{V_p V_e}} e^{\frac{\pi}{2\kappa}} \Gamma\left(1 - \frac{i}{\kappa}\right) \frac{1}{\sqrt{V_e}} e^{\frac{\pi}{2k'_e}} \Gamma\left(1 - \frac{i}{k'_e}\right) \Big[k'_e (\nabla_{k'_e} \mathcal{J})$$
$$+ k'_e (\nabla_{\boldsymbol{q}} \mathcal{J}) + i\boldsymbol{k}'_e \left(\frac{\partial}{\partial\lambda} \mathcal{J}\right) \Big]. \tag{6.30}$$

Here, the derivative $\frac{\partial}{\partial\lambda}\mathcal{J}$ is given by (6.22). Further, taking into consideration the solution of the integral $\mathcal{J}$ from (6.16) and the differentiation formula (see, e.g. [87]) $\frac{d}{dz}F(a, b, c, z) = \frac{ab}{c}F(a+1, b+1, c+1, z)$ for the hypergeometric function, the gradients $\nabla_{k'_e}\mathcal{J}$ and $\nabla_{\boldsymbol{q}}\mathcal{J}$ are easily calculated. They read

$$\nabla_{k'_e} \mathcal{J} = \frac{2\pi}{\tilde{\alpha}} e^{-\frac{\pi}{\kappa}} \left(\frac{\tilde{\alpha}}{\tilde{\gamma}}\right)^{\frac{i}{\kappa}} \left(\frac{\tilde{\gamma}+\tilde{\delta}}{\tilde{\gamma}}\right)^{-i\nu} \Big[\boldsymbol{A}_3 F\left(1 - \frac{i}{\kappa}, i\nu, 1, z\right)$$
$$+ \boldsymbol{A}_4 F\left(2 - \frac{i}{\kappa}, i\nu + 1, 2, z\right) \Big] \tag{6.31}$$

and

$$\nabla_{\boldsymbol{q}} \mathcal{J} = \frac{2\pi}{\tilde{\alpha}} e^{-\frac{\pi}{\kappa}} \left(\frac{\tilde{\alpha}}{\tilde{\gamma}}\right)^{\frac{i}{\kappa}} \left(\frac{\tilde{\gamma}+\tilde{\delta}}{\tilde{\gamma}}\right)^{-i\nu} \Big[\boldsymbol{A}_5 F\left(1 - \frac{i}{\kappa}, i\nu, 1, z\right)$$
$$+ \boldsymbol{A}_6 F\left(2 - \frac{i}{\kappa}, i\nu + 1, 2, z\right) \Big] \tag{6.32}$$

with the rather cumbersome vectors $\boldsymbol{A}_{3,4,5,6} = \boldsymbol{A}_{3,4,5,6}(\boldsymbol{k}_p, \boldsymbol{k}_e, \boldsymbol{k}'_e)$ specified in Appendix 9.5 in the Electronic Supplementary Material. Note that $\nabla_{\boldsymbol{k}'_e}\mathcal{J}$ was evaluated under the assumption $\nu = 1/k'_e = \text{const.}$ with respect to $\nabla_{\boldsymbol{k}'_e}$.

Inserting (6.22), (6.31) and (6.32) into (6.30) and afterwards inserting the resulting expression for $\mathcal{I}_5$ as well as the result for $\mathcal{I}_4$ from (6.26) into (6.25), the quantity $\mathcal{M}_3$ yields

$$\mathcal{M}_3 = 8\pi^2 iN \frac{\beta^2-\alpha^2}{(\alpha^2+P^2)(\beta^2+P^2)} \frac{1}{\sqrt{V_p V_e}} e^{-\frac{\pi}{2\kappa}} \Gamma\left(1-\frac{i}{\kappa}\right) \frac{1}{\sqrt{V_e}} e^{\frac{\pi}{2k'_e}} \Gamma\left(1-\frac{i}{k'_e}\right) k'_e$$
$$\times \frac{1}{\tilde{\alpha}}\left(\frac{\tilde{\alpha}}{\tilde{\gamma}}\right)^{\frac{i}{\kappa}} \left(\frac{\tilde{\gamma}+\tilde{\delta}}{\tilde{\gamma}}\right)^{-i\nu} \left[\boldsymbol{P}\cdot\left(i\frac{\boldsymbol{k}'_e}{k'_e}A_1 + \boldsymbol{A}_3 + \boldsymbol{A}_5\right) F\left(1-\frac{i}{\kappa}, i\nu, 1, z\right)\right.$$
$$\left. + \boldsymbol{P}\cdot\left(i\frac{\boldsymbol{k}'_e}{k'_e}A_2 + \boldsymbol{A}_4 + \boldsymbol{A}_6\right) F\left(2-\frac{i}{\kappa}, i\nu+1, 2, z\right)\right]. \tag{6.33}$$

We now apply the results for $\mathcal{M}_{1-3}$ from (6.17), (6.23) and (6.33) in order to obtain the final expression for the transition amplitude (6.9), which is given by

$$a_{3BAe} = \frac{16\pi^3 N(\beta^2-\alpha^2)}{i\sqrt{V_p}V_e(\alpha^2+P^2)(\beta^2+P^2)} e^{-\frac{\pi}{2\kappa}} \Gamma\left(1-\frac{i}{\kappa}\right) e^{\frac{\pi}{2k'_e}} \Gamma\left(1-\frac{i}{k'_e}\right)$$
$$\times\, Q_{pe}\,\delta(\varepsilon_{k_p} - \varepsilon_b - (\varepsilon_{k'_e} - \varepsilon_{k_e})), \tag{6.34}$$

where

$$Q_{pe} = \frac{1}{\tilde{\alpha}}\left(\frac{\tilde{\alpha}}{\tilde{\gamma}}\right)^{\frac{i}{\kappa}} \left(\frac{\tilde{\gamma}+\tilde{\delta}}{\tilde{\gamma}}\right)^{-\frac{i}{k'_e}} \left[\Lambda_1 F\left(1-\frac{i}{\kappa}, \frac{i}{k'_e}, 1, z\right)\right.$$
$$\left. -\Lambda_2 F\left(2-\frac{i}{\kappa}, \frac{i}{k'_e}+1, 2, z\right)\right] \tag{6.35}$$

with

$$\Lambda_1 = 1 - \varepsilon_{k'_e} A_1 - ik'_e \boldsymbol{P}\cdot\left(i\frac{\boldsymbol{k}'_e}{k'_e}A_1 + \boldsymbol{A}_3 + \boldsymbol{A}_5\right),$$
$$\Lambda_2 = \varepsilon_{k'_e} A_2 + ik'_e \boldsymbol{P}\cdot\left(i\frac{\boldsymbol{k}'_e}{k'_e}A_2 + \boldsymbol{A}_4 + \boldsymbol{A}_6\right). \tag{6.36}$$

The formation rate of $\bar{\mathrm{H}}^+$ ions per unit of time (per $\bar{\mathrm{H}}$) for electron-assisted three-body attachment is determined by calculating the quantity

$$\mathcal{R}_{3BAe} = \frac{V_e}{(2\pi)^3} \int d^3k'_e \, \frac{|a_{3BAe}|^2}{T}. \tag{6.37}$$

In (6.37), we integrate over the momentum $\boldsymbol{k}'_e$ of the outgoing electron and, similar as for the spontaneous radiative attachment rate (5.14), the time duration T is of the order of the interaction time. Inserting the amplitude (6.34) into (6.37) and employing the identity $\left[2\pi\delta(\varepsilon_{k_p}-\varepsilon_b-(\varepsilon_{k'_e}-\varepsilon_{k_e}))\right]^2 = 2\pi\delta(\varepsilon_{k_p}-\varepsilon_b-(\varepsilon_{k'_e}-\varepsilon_{k_e}))T$, the rate $\mathcal{R}_{3BAe}$ becomes

$$\begin{aligned}\mathcal{R}_{3BAe} = {} & \frac{16\pi^2 N^2(\beta^2-\alpha^2)^2}{V_p V_e(\alpha^2+P^2)^2(\beta^2+P^2)^2} e^{-\frac{\pi}{\kappa}} \left|\Gamma\left(1-\frac{i}{\kappa}\right)\right|^2 \\ & \times \int d^3k'_e \, \delta(\varepsilon_{k_p}-\varepsilon_b-(\varepsilon_{k'_e}-\varepsilon_{k_e})) e^{\frac{\pi}{k'_e}} \left|\Gamma\left(1-\frac{i}{k'_e}\right)\right|^2 |Q_{pe}|^2.\end{aligned} \tag{6.38}$$

We use the relation $\delta(g(x)) = \sum_{x_j \in \{\text{simple roots of } g(x)\}} \frac{1}{|g'(x_j)|}\delta(x-x_j)$ and write the delta function in (6.38) as $\delta(\varepsilon_{k_p}-\varepsilon_b-(\varepsilon_{k'_e}-\varepsilon_{k_e})) = \frac{1}{k'_c}\delta(k'_e-k'_c)$, where $k'_c = \sqrt{k_p^2+k_e^2-2\varepsilon_b}$. Subsequently, performing the integration over the absolute value $k'_e = |\boldsymbol{k}'_e|$ of the outgoing electron momentum and expressing the positron and electron normalization volumes V_p and V_e by the corresponding number densities n_p and n_e of positrons and electrons according to $V_p = 1/n_p$ and $V_e = 1/n_e$, the formation rate $\mathcal{R}_{3BAe}$ is obtained to be

$$\mathcal{R}_{3BAe} = \frac{16\pi^2 n_p n_e N^2(\beta^2-\alpha^2)^2}{(\alpha^2+P^2)^2(\beta^2+P^2)^2} e^{-\frac{2\pi}{\kappa}} \, G(\kappa)\, G(k'_c) k'_c \int d\Omega_{k'_e} (|Q_{pe}|^2)_{k'_e=k'_c}. \tag{6.39}$$

Here,

$$G(k) = e^{\frac{\pi}{k}} \left|\Gamma\left(1-\frac{i}{k}\right)\right|^2 = \frac{2\pi}{k(1-e^{-\frac{2\pi}{k}})} \tag{6.40}$$

is the Gamow factor for a charged particle with absolute momentum k that moves in an attractive Coulomb field (see, e.g. [137]).

In the following, we consider antihydrogen atoms embedded in a gas of low energy ($\approx$ meV) electrons which is penetrated by a positron beam. We assume that the beam of positrons propagates in a fixed direction and its (sharp) energy varies in a relatively broad range from sub-meV up to eV. In such a case, the $\bar{\mathrm{H}}^+$ formation rate can be obtained by averaging (6.39) over the absolute value k_e of the incident electron momentum by applying a Maxwell-Boltzmann distribution. Furthermore, we account for all relative orientations between the incident positron and the incident electron momenta by performing the average over the solid angle Ω_{k_e} of the incident electron momentum while the direction of the incident positron is fixed. Therefore, the averaged formation rate $\langle \mathcal{R}_{3BAe} \rangle$ of $\bar{\mathrm{H}}^+$ ions per unit of time (per $\bar{\mathrm{H}}$) can be calculated according to

$$\langle \mathcal{R}_{3BAe} \rangle = \frac{1}{4\pi} \int d\Omega_{k_e} \int_0^\infty dk_e \, w_{E_T}(k_e) \mathcal{R}_{3BAe}, \tag{6.41}$$

where $w_{E_T}(k_e) = \frac{4\pi k_e^2}{(2\pi E_T)^{3/2}} e^{-\frac{k_e^2}{2E_T}}$ is the Maxwell-Boltzmann distribution for the incident electron. In addition, $E_T = k_B T$ is the average thermal energy associated with the electron gas of temperature T, where k_B is the Boltzmann constant. It is worth mentioning that the rate (6.41) now only depends on the absolute momentum k_p of the incident positron.

6.2 Positron-assisted Three-body Attachment

We suppose a situation, where free positrons e^+ move in close proximity to antihydrogen atoms $\bar{\mathrm{H}}$. Then, one positron may be attached to $\bar{\mathrm{H}}$ – driven by the positron-positron interaction – whereas another positron carries away the energy release. A scheme of this process, referred to as positron-assisted three-body attachment, can be found in Fig. 6.2, where $\varepsilon_{\tilde{k}_p}$ and $\varepsilon_{\tilde{k}'_p}$ are the energies of the incident and outgoing assisting positron, respectively.

The difference between electron-assisted and positron-assisted three-body attachment is that the incident electron involved in the former process is replaced by a second incident positron in the latter process. Hence, their theoretical treatments are quite similar but with two changes. The first minor change is that the positron-positron interaction in the treatment of 3BAp has a different sign than the positron-electron interaction in the 3BAe treatment. The second major change involves the Coulomb wave functions appearing in the initial and final states of the effectively three-body system. In particular, the Coulomb wave (6.4) which accounts for the attractive Coulomb interaction between the incident electron and positron in the initial state in the treatment of 3BAe has to be replaced by a corresponding Coulomb wave that takes into account the repulsive Coulomb interaction between

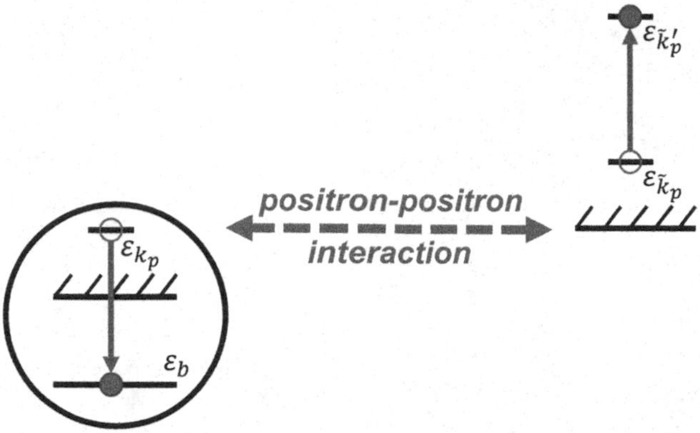

$$e^+ + e^+ + \overline{\mathrm{H}} \rightarrow e^+ + \overline{\mathrm{H}}^+$$

Fig. 6.2 Scheme of positron-assisted three-body attachment (3BAp). (This figure was originally published in Ref. [120])

the two incident positrons in the initial state in the 3BAp treatment. Accordingly, the Coulomb wave (6.6) which accounts for the attractive Coulomb interaction between the outgoing electron and the attached positron in the final state in the treatment of 3BAe has to be replaced by a corresponding Coulomb wave that takes into account the repulsive Coulomb interaction between the outgoing positron and the attached positron in the final state in the 3BAp treatment.

Accounting for the above changes in the theoretical treatment in Section 6.1 results in a formation rate of $\bar{\mathrm{H}}^+$ ions for 3BAp that is many orders of magnitude smaller than the rate for 3BAe, which will be discussed in detail later in Section 7.1.3.

Numerical Results and Discussion 7

This chapter provides a comparative discussion of the numerical results for the radiative and nonradiative attachment mechanisms under consideration. In this context, we also briefly discuss the efficiency of positron capture in collisions of positronium with antihydrogen. Further, we discuss the role of annihilation and other processes related to the interaction between matter and antimatter for two-center dileptonic and electron-assisted three-body attachment, both in which matter particles are involved. The following chapter is mainly based on results published initially in Refs. [120, 121].

7.1 Comparative Analysis of the Radiative and Nonradiative Attachment Mechanisms

The $\bar{\mathrm{H}}^+$ formation rates for all considered attachment mechanisms depend on the density n_p of incident positrons. Here, we choose $n_p = 10^8$ cm^{-3}, which corresponds to the typical positron density in $\bar{\mathrm{H}}$ and $\bar{\mathrm{H}}^+$ experiments (see, e.g. [99, 138]).

7.1.1 Spontaneous Radiative vs. Two-Center Dileptonic vs. Electron-assisted Three-body Attachment

For evaluating the two-center dileptonic attachment rate (5.48), we consider the case when positrons and a beam of slow $\bar{\mathrm{H}}$ move in a gas of Cs atoms. The corresponding attachment reaction can be written as $(\bar{\mathrm{H}}+\mathrm{e}^+)+\mathrm{Cs}(6\,^2S_{1/2}) \rightarrow \bar{\mathrm{H}}^+ + \mathrm{Cs}(6\,^2P_{3/2}) \rightarrow \bar{\mathrm{H}}^+ + \mathrm{Cs}(6\,^2S_{1/2}) + \hbar\omega_k$. Consequently, the positron capture by antihydrogen

A. Jacob, *Relativistic Effects in Interatomic Ionization Processes and Formation of Antimatter Ions in Interatomic Attachment Reactions*,
https://doi.org/10.1007/978-3-658-43891-3_7

proceeds due to the transfer of the energy release to Cs, exciting the $6\,^2S_{1/2} \rightarrow 6\,^2P_{3/2}$ dipole transition ($\omega_B \approx 1.46$ eV) with subsequent de-excitation via spontaneous radiative decay ($\Gamma_r^B \approx 3.41 \times 10^{-9}$ eV [56]). Based on the discussion of collisional two-center dielectronic recombination (attachment) in [41, 135], we choose here $b_{min} = 5$ a.u. in order to account for as much of the total rate as possible while at the same time satisfying all assumptions which the approach to the two-center process relies on. In addition, we set $n_B = 10^{15}$ cm^{-3}, $v = 0.01$ a.u. and $\vartheta_{k_p} = \pi/2$. Note that averaging the rate (5.48) over the direction of the incident positron yields a result which is ≈ 34 % smaller than that at a fixed $\vartheta_{k_p} = \pi/2$.

A numerical result for the electron-assisted three-body attachment rate (6.41) is obtained by performing the average over the absolute value of the momentum of the incident electron, using a Maxwell-Boltzmann distribution with an average thermal energy $E_T = k_B T = 1$ meV (≈ 11.6 K). Such energy is in the range of typical energies for electrons or positrons in the cryogenic environments of CERN experiments on $\bar{\text{H}}$ and $\bar{\text{H}}^+$ (see, e.g. [99, 138]). Further, we choose the incident electron density as $n_e = 5 \times 10^{10}$ cm^{-3} which, at the moment, is the highest possible density of electrons that can be experimentally realized in a cryogenic environment at temperatures $T \approx 10$ K [139].

In Fig. 7.1, we illustrate the rates (5.17), (5.48) and (6.41) for the formation of $\bar{\text{H}}^+$ ions via spontaneous radiative attachment (the dotted curve), two-center dileptonic attachment (the dashed curve) and electron-assisted three-body attachment (the solid curve) as a function of the incident positron energy.

We can conclude from Fig. 7.1 that, in the interval of energies of the incident positron from 10^{-4} to 2 eV, the rate $\langle \mathcal{R}_{3BAe} \rangle$ is maximal at the smallest energy displayed, changes only weakly between 10^{-4} and 10^{-3} eV and then decreases faster and faster with further increasing energy. The dependence of the rate $\langle \mathcal{R}_{3BAe} \rangle$ on the incident positron energy reflects the presence of the Coulomb singularity in the wave function (6.4), describing the relative motion of the incident electron-positron pair, as the energy of this motion approaches 0. Consequently, the observed decrease of the rate $\langle \mathcal{R}_{3BAe} \rangle$ results from a corresponding decrease in the strength of the Coulomb interaction between the electron and positron when their relative energy increases.

The rate $\mathcal{R}_{SRA}$ behaves rather differently. It grows when ε_{k_p} increases and subsequently saturates at the largest energies shown (and decreases for even higher energies). In contrast to the long-range Coulomb interaction between the incident electron and positron present in the 3BAe process, here the interaction of the incident positron with a neutral $\bar{\text{H}}(1s)$ is of short range and the rate $\mathcal{R}_{SRA}$ has a minimum (zero) as the incident positron energy tends to 0. In fact, within the dipole approximation, the incident positron must be in a state with one unit of the orbital angular

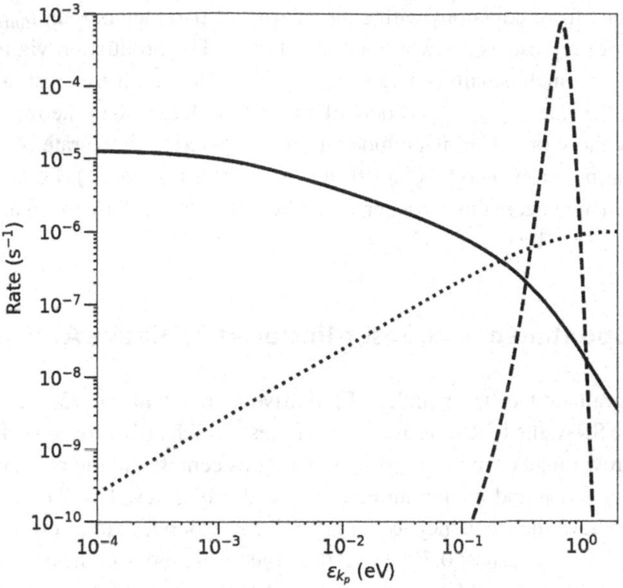

Fig. 7.1 The $\bar{\mathrm{H}}^+$ formation rate per unit of time (per $\bar{\mathrm{H}}$) as a function of the energy ε_{k_p} of the incident positron for spontaneous radiative attachment (dotted), two-center dileptonic attachment (dashed) and electron-assisted three-body attachment (solid). See text for the choice of parameters. (This figure was originally published in Ref. [121])

momentum, so there is a centrifugal barrier (leading to an effective repulsive force), which at very low energies does not allow the positron to come close enough to $\bar{\mathrm{H}}(1s)$ that is necessary for the SRA to occur.

In addition, the rate $\mathcal{R}_{2CDA}$ exhibits a resonant shape, having a maximum close to the resonant positron energy at $\varepsilon_{k_{p,res}} = \varepsilon_b + \omega_B \approx 0.71$ eV and decreasing rapidly when deviating from this point. Due to the relative motion of $\bar{\mathrm{H}}$ and Cs, the width of the maximum ($\delta\varepsilon_{k_p} \sim v/b_{min} = 2\times10^{-3}$ eV) is rather large and it exceeds the corresponding radiative width of the excited state of Cs ($\Gamma_r^B \approx 3.41 \times 10^{-9}$ eV) by several orders of magnitude.

It can be seen in Fig. 7.1 that the 2CDA is only competitive for energies of the incident positron which are close to the resonance energy at $\varepsilon_{k_{p,res}} \approx 0.71$ eV. Especially, exactly on the resonance, the 2CDA is much more efficient than the SRA and 3BAe, dominating the SRA by a factor $\approx 9.3 \times 10^2$ and the 3BAe by a factor $\approx 4.9 \times 10^3$. (We recall that the rate $\mathcal{R}_{2CDA}$ does not account for contributions from collisions with smaller impact parameters than $b_{min} = 5$ a.u. and therefore represents a lower boundary for the $\bar{\mathrm{H}}^+$ production via the 2CDA.) However, for (much) smaller energies $\varepsilon_{k_p} \lesssim 10^{-2}$ eV, which are most favorable for the 3BAe, the rate $\langle \mathcal{R}_{3BAe} \rangle$ is orders of magnitude larger than the rate $\mathcal{R}_{SRA}$ and even many more orders of magnitude larger compared with the rate $\mathcal{R}_{2CDA}$ (since the incident positron energy is far off the two-center resonance). Furthermore, the 3BAe mechanism remains stronger than the SRA and 2CDA mechanisms up to energies $\varepsilon_{k_p} \approx 0.1$ eV.

7.1.2 Spontaneous vs. (laser-)induced Radiative Attachment

Now, we evaluate the (laser-)induced radiative attachment rate (5.47). We remind that, unlike SRA, the LIRA is a resonant process which only proceeds efficiently in a very narrow range of incident positron energies centered at the resonance energy $\varepsilon_{k_{p,res}} = \varepsilon_b + \omega_0$ and having an effective width of a few Γ's. The laser field is assumed to have the frequency $\omega_0 = 1.5$ eV, corresponding to a resonant positron energy $\varepsilon_{k_{p,res}} = \varepsilon_b + \omega_0 \approx 0.75$ eV. Further, the (averaged over the period) intensity $I_0 = cF_0^2/(8\pi)$ of the field is chosen as $I_0 = 10^6$ W/cm^2, which is sufficiently weak so that the laser field does not destroy the produced $\bar{\mathrm{H}}^+$ ions (see Section 5.2). In addition, we set $\Delta\varepsilon_{k_p} = 0.1$ eV and assume that the duration T of the laser pulse is not too long, $\Gamma t \leq \Gamma T \ll 1$, such that $e^{-\Gamma t} \approx 1$ (note that $\Gamma \approx 2 \times 10^7$ s^{-1} at $I_0 = 10^6$ W/cm^2). Then, the rate (5.47) for the formation of $\bar{\mathrm{H}}^+$ ions via the LIRA is obtained to be $\langle \mathcal{R}_{LIRA} \rangle \approx 2.1 \times 10^{-5}$ s^{-1}. It outperforms the corresponding SRA rate, evaluated at $\varepsilon_{k_p} = 0.75$ eV, by a factor ≈ 26.

It is worth mentioning that there were suggestions ([140, 141], see also [99]) to increase the production of antihydrogen atoms $\bar{\mathrm{H}}$ by using (laser-)induced recombination of positrons with antiprotons. However, to our knowledge, there is no experimental evidence for this process in collisions between positrons and antiprotons. Especially, the induced recombination could not be confirmed in an experiment [142] in which no effect of the laser field on the antihydrogen formation was observed. This was explained by the dominance of three-body recombination, $\mathrm{e}^+ + \mathrm{e}^+ + \bar{\mathrm{p}} \rightarrow \mathrm{e}^+ + \bar{\mathrm{H}}$, under the given experimental conditions, where the capture of low energy positrons into highly excited Rydberg states of $\bar{\mathrm{H}}$ was most probable. Note that such states are absent for the formation of the $\bar{\mathrm{H}}^+$ ion. Moreover, as we will see below, the rate for $\bar{\mathrm{H}}^+$ formation in collisions between positrons and antihydrogen atoms via three-body attachment, $\mathrm{e}^+ + \mathrm{e}^+ + \bar{\mathrm{H}} \rightarrow \mathrm{e}^+ + \bar{\mathrm{H}}^+$, is vanishingly small for all positron energies under consideration. Therefore, the (laser-)induced radiative attachment of a positron to antihydrogen is not expected to be hidden by the nonradiative three-body attachment.

7.1.3 Electron-assisted vs. Positron-assisted Three-body Attachment

Next, let us briefly consider positron-assisted three-body attachment, where the corresponding formation rate $\langle \mathcal{R}_{3BAp} \rangle$ is evaluated similarly to the rate $\langle \mathcal{R}_{3BAe} \rangle$ for electron-assisted three-body attachment. In Fig. 7.2, we show the rate $\langle \mathcal{R}_{3BAp} \rangle$ as a function of the incident positron energy and, for a comparison, we also display the rate $\langle \mathcal{R}_{3BAe} \rangle$ from Fig. 7.1.

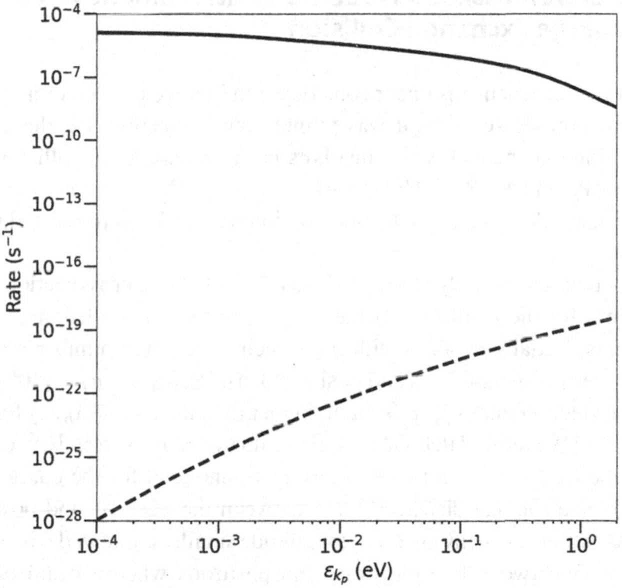

Fig. 7.2 The $\bar{\mathrm{H}}^+$ formation rate per unit of time (per $\bar{\mathrm{H}}$) as a function of the energy ε_{k_p} of the incident positron for electron-assisted (solid) and positron-assisted (dashed) three-body attachment

It can be observed in Fig. 7.2 that, in the range of positron energies from 10^{-4} to 2 eV, the 3BAp mechanism has vanishingly small formation rates as compared with the 3BAe mechanism. We can explain this result by the fact that an attraction between the incident positron and electron in the 3BAe is replaced by a repulsion between two incident positrons in the 3BAp, not allowing them to come close to each other, which (strongly) weakens the attachment reaction.

Note however that the corresponding three-body recombination reaction leading to the formation of an antihydrogen atom, $e^+ + e^+ + \bar{p} \rightarrow e^+ + \bar{H}$, can be rather efficient. It is indeed used for the production of antihydrogen in laboratories [97–99]. As compared with the formation of $\bar{H}^+$ via positron-assisted three-body attachment, here the mutual repulsion of the two incident positrons is compensated by their attraction to the antiproton, which is now not screened by the bound (anti)atomic positron.

7.1.4 Electron-assisted Three-body Attachment vs. Ps $-$ $\bar{H}$ Charge Exchange Collision

The attachment mechanisms under consideration involve capture of a free positron by antihydrogen. However, as it was pointed out in Section 4.2, there exist another $\bar{H}^+$ formation channel which involves positron capture in collisions between positronium Ps and $\bar{H}$: $\mathrm{Ps} + \bar{H}(1s) \rightarrow e^- + \bar{H}^+$. For the sake of completeness, we compare this charge exchange reaction to the electron-assisted three-body attachment.

A recent theoretical study [143] predicts relatively large cross sections ($\sim 10^{-16}$ - 10^{-15} cm^2) for the formation of the $\bar{H}^+$ ion in Ps $-$ $\bar{H}$ collisions, where Ps is assumed to be initially in states with the principal quantum number $n \in \{1, 2, 3\}$. In particular, for comparable Ps and positron densities ($n_{\mathrm{Ps}} \approx n_p = 10^8$ cm^{-3}), the resulting production rate $\mathcal{R}_{\mathrm{Ps}-\bar{H}}$ is much higher than the rate $\langle \mathcal{R}_{3BAe} \rangle$ for the 3BAe from Fig. 7.1. (We remind that $\langle \mathcal{R}_{3BAe} \rangle$ is evaluated at $n_e = 5 \times 10^{10}$ cm^{-3}.) This can be explained by the fact that it is strongly beneficial for the charge exchange reaction to have average distances $\langle a_{\mathrm{Ps}} \rangle$ between the electron and positron in Ps bound states which are many orders of magnitude smaller compared with the average distances $\langle r_{ep} \rangle$ between free electrons and positrons when a beam of positrons penetrates an electron gas. More precisely, in the ground state of Ps the average distance is $\langle a_{\mathrm{Ps}} \rangle \approx 1.1 \times 10^{-8}$ cm (see, e.g. [144]) while for a gas of electrons having the density $n_e = 5 \times 10^{10}$ cm^{-3} the average distance is obtained to be $\langle r_{ep} \rangle = n_e^{-1/3} \approx 2.7 \times 10^{-4}$ cm. This huge difference even overcompensates a much weaker intrinsic capture efficiency for an initially bound positron compared to an initially free positron.

Considering Ps which is initially in a state with the principal quantum number $n = 3$, the energy threshold for $\bar{\text{H}}^+$ formation in $\text{Ps} - \bar{\text{H}}$ collisions is ≈ 1.7 meV [143], which is in the range of the very low relative energies where the 3BAe is most efficient. In this case, taking advantage of the results in [143], we obtain $\mathcal{R}_{\text{Ps}-\bar{\text{H}}} \approx \langle \mathcal{R}_{3BAe} \rangle$ starting already at $n_e \approx 2 \times 10^{14}$ cm^{-3} for which $\langle r_{ep} \rangle \approx 1.7 \times 10^{-5}$ cm that is still much larger than $\langle a_{\text{Ps}} \rangle$.

Note however that if Ps is initially in the ground state (in states having the principal quantum number $n = 2$), the energy threshold for $\bar{\text{H}}^+$ production in $\text{Ps} - \bar{\text{H}}$ collisions is ≈ 6.05 eV (≈ 0.95 eV) [143]. Consequently, in such a case, the charge exchange reaction cannot compete with the 3BAe process at the very low relative energies at which the 3BAe is most efficient.

7.2 The Role of Annihilation and other Processes Related to the Interaction between Matter and Antimatter

The results in Section 7.1 suggests that the $\bar{\text{H}}^+$ formation via two-center dileptonic and electron-assisted three-body attachment can be quite efficient. However, the 2CDA and 3BAe mechanisms proceed in environments consisting of matter and antimatter and hence the question naturally arises whether annihilation and other processes related to the interaction between matter and antimatter would not effectively eliminate them.

7.2.1 Two-Center Dileptonic Attachment in an Environment Consisting of Matter and Antimatter

The particles to be considered in an environment where two-center dileptonic attachment takes place are free positrons, $\bar{\text{H}}$ atoms, $\bar{\text{H}}^+$ ions (that are produced) and neutral atoms B. Thus, we have to discuss what can happen to a free positron and to $\bar{\text{H}}$ and $\bar{\text{H}}^+$ which move in a gas of neutral atoms B.

Let us first suppose a positron which penetrates a gas of neutral atoms. In this case, elastic positron scattering (see, e.g. [145] and references therein) is by far the dominant process. Especially for the incident positron energies of interest ($\varepsilon_{k_p} \lesssim 1$ eV) its cross section is rather large (up to a few tens of 10^{-16} cm^2). Yet, elastic scattering does not affect the number and energy of free positrons. Therefore, this process is not expected to significantly impact the efficiency of the 2CDA.

The cross sections for positron impact excitation of neutral atoms can be quite substantial. However, at the (relatively low) positron energies under consideration, excitation from the ground state of neutral atoms is not allowed by the energy conservation.

If a positron moves in the close vicinity of a neutral atom, positronium Ps can be formed via a charge exchange collision. In case of collisions between positrons and Cs atoms the cross section for Ps formation at the positron energies of interest is $\sigma_{\mathrm{Ps}} \approx 2 \times 10^{-16}$ cm^2 [146]. The mean free path l_p of a positron in a gas of Cs atoms with regard to this process is given by $l_p = (\sigma_{\mathrm{Ps}} n_B)^{-1} \approx 5 \times 10^3$ cm and ≈ 5 cm at an atomic density of $n_B \approx 10^{12}$ cm^{-3} and $\approx 10^{15}$ cm^{-3}, respectively. Consequently, although the charge exchange collision reduces the total number of positrons which are available for the 2CDA mechanism, this process is not assumed to crucially impact the efficiency of the 2CDA (unless the density of atoms B approaches rather high values). Moreover, it is worth mentioning that the formation of Ps does not eliminate the pathway for the production of $\bar{\mathrm{H}}^+$ ions since the latter can still be formed via the charge exchange collision $\mathrm{Ps} + \bar{\mathrm{H}} \rightarrow e^- + \bar{\mathrm{H}}^+$ (see Section 7.1.4).

In addition, a positron may form a bound state with some neutral atoms, which would reduce the total number of positrons available for the 2CDA process. Yet, we note that it is unlikely for a positron to form a bound state with Cs (or Rb) atoms [147].

The last process to consider when a positron moves in a gas of neutral atoms is positron annihilation. In contrast to the previous processes, positron annihilation would completely terminate the formation of $\bar{\mathrm{H}}^+$ ions. However, annihilation of a positron with a bound atomic electron is relatively unlikely. In particular, if we assume that a positron annihilates mainly with electrons from outer atomic shells by two-photon emission and that this process can essentially be considered as annihilation of a free positron-electron pair, the annihilation cross section at the positron energies of interest is $\sigma_{annihil.} \approx 1.7 \times 10^{-22}$ cm^2, which is very small. The value for $\sigma_{annihil.}$ was obtained by using formulas for the annihilation cross section of a free positron-electron pair from [148]. At an atomic density of $n_B \approx 10^{15}$ cm^{-3}, the mean free path of a positron with respect to the annihilation process is very large, $l_p = (\sigma_{annihil.} n_B)^{-1} \approx 59$ km. Therefore, we can expect that positron annihilation will not have any noticeable impact on the efficiency of the 2CDA.

Besides from free positrons, we should also consider $\bar{\mathrm{H}}$ and $\bar{\mathrm{H}}^+$ moving in a gas of neutral atoms B. To our knowledge there exist neither experimental nor theoretical studies for processes which involve $\bar{\mathrm{H}}^+$ ions penetrating matter. However, at the impact energies of interest ($\approx 10 - 150$ eV/u), there exist theoretical results for collisions between the $\bar{\mathrm{H}}$ atom and the simplest matter atoms and molecules (H, He, H_2, H_2^+) (see [149]). On their basis, we might expect that in collisions with impact energies $\gtrsim 20 - 30$ eV/u annihilation of antiprotons will not be the main reason for $\bar{\mathrm{H}}$ and $\bar{\mathrm{H}}^+$ losses.

Furthermore, since the $\bar{\mathrm{H}}^+$ ion has a much lower binding energy and a much bigger size ($r_{\bar{\mathrm{H}}^+} \approx 4.26$ a.u.) than $\bar{\mathrm{H}}$, it is reasonable to suppose that the loss of $\bar{\mathrm{H}}^+$ in collisions with matter atoms will be significantly larger compared with the loss of $\bar{\mathrm{H}}$. In order to have at least some rough estimate for the $\bar{\mathrm{H}}^+$ loss, we assume that any collision between $\bar{\mathrm{H}}^+$ and a matter atom that occurs in the range of impact parameters $0 \leq b \leq r_{\bar{\mathrm{H}}^+}$ will result in the destruction of the $\bar{\mathrm{H}}^+$ ion (for one reason or another). Then, the corresponding total loss cross section is given by $\sigma_{loss} = \pi r^2_{\bar{\mathrm{H}}^+} \approx 1.6 \times 10^{-15}$ cm^2 and the mean-free path with respect to the loss of $\bar{\mathrm{H}}^+$ is obtained to be $l_{\bar{\mathrm{H}}^+} = (\sigma_{loss} n_B)^{-1} \approx 6.3 \times 10^2$ cm and ≈ 0.63 cm at an atomic density of $n_B \approx 10^{12}$ cm^{-3} and $\approx 10^{15}$ cm^{-3}, respectively.

7.2.2 Electron-assisted Three-body Attachment in an Environment Consisting of Matter and Antimatter

The particles to be taken into account in an environment in which electron-assisted three-body attachment occurs are free positrons and electrons, $\bar{\mathrm{H}}$ atoms as well as $\bar{\mathrm{H}}^+$ ions (that are produced). We point out that the positron-electron annihilation, proceeding either in a free positron-electron pair or in a pair consisting of a bound positron and free electron, is the only process which possibly affects the efficiency of the 3BAe mechanism.

We first consider annihilation of a free positron-electron pair at an energy of 1 meV for the relative motion between the positron and electron. Applying formulas for the annihilation cross section of a free positron-electron pair from [148], we arrive at a quite small cross section $\sigma_{annihil.} \approx 4 \times 10^{-21}$ cm^2. Consequently, the mean-free path of positrons in a gas of electrons (having the density $n_e = 5 \times 10^{10}$ cm^{-3}) with respect to the annihilation process is obtained to be $l_p = (\sigma_{annihil.} n_e)^{-1} \approx 5 \times 10^7$ m, which is huge. Thus, annihilation of a free positron-electron pair is not assumed to have any noticeable impact on the efficiency of the 3BAe. Note that a free positron-electron pair could also emit a photon to form a positronium which is eventually annihilated, but at the low energies under consideration the cross section for this process is very small and can be neglected.

When free electrons pass close by $\bar{\mathrm{H}}$, annihilation in a subsystem of a bound positron and free electron can occur. Supposing that the relative velocity between the electron and positron is $\sim$ 1 a.u. and employing formulas from [148], the resulting annihilation cross section, $\sigma_{annihil.} \sim 10^{-23}$ cm^2, is even smaller than the cross section for annihilation in a free positron-electron pair.

It is worth mentioning that, in case of free electrons passing close by $\bar{\mathrm{H}}$, one could also imagine the formation of positronium via positron capture from $\bar{\mathrm{H}}$ by a free electron or the existence of bound or long-lived resonance states in the $e^- - \bar{\mathrm{H}}$ system. However, at the energies of interest, positronium formation is energetically forbidden and since it is known that the $e^+ - \mathrm{H}$ system neither has stable bound states nor has, at low positron energies, long-lived resonance states (see, e.g. [150]), we can expect all corresponding states in the $e^- - \bar{\mathrm{H}}$ system to be absent as well.

Summary and Concluding Remarks

8

We have considered the formation of the positive ion of antihydrogen via radiative and nonradiative attachment of an incident positron to an antihydrogen atom.

Three radiative formation mechanisms were discussed, all of which have photoemission as a common feature. The first mechanism is spontaneous radiative attachment of an incident positron to antihydrogen, $e^+ + \bar{H} \rightarrow \bar{H}^+ + \hbar\omega_k$. The driving force of this mechanism is the interaction of the $e^+ - \bar{H}$ system with the (quantum) radiation field resulting in spontaneous emission of a photon which takes away the energy excess in the attachment process. The second mechanism is (laser-)induced radiative attachment, $e^+ + \bar{H} + N\hbar\omega_0 \rightarrow \bar{H}^+ + (N+1)\hbar\omega_0$. This mechanism is driven by the interaction of the $e^+ - \bar{H}$ system with a relatively weak laser field, resonantly tuned to positron transitions, leading to induced photoemission. The last radiative mechanism is two-center dileptonic attachment, $e^+ + \bar{H} + B \rightarrow \bar{H}^+ + B^* \rightarrow \bar{H}^+ + B + \hbar\omega_k$, which proceeds when beams of positrons and antihydrogens cross in a gas of (matter) atoms B. Then, the attachment takes place by resonant transfer of energy from the $e^+ - \bar{H}$ subsystem to atom B via the two-center positron-electron (dileptonic) interaction, resulting in excitation of B with its subsequent relaxation through spontaneous radiative decay. Therefore, similar to (laser-)induced radiative attachment, two-center dileptonic attachment is a resonant process (although its resonance nature is strongly washed out by the relative motion of $\bar{H}$ and B). Furthermore, in contrast to the other radiative mechanisms, the two-center process involves two interactions, namely the (Coulomb) interaction between the $e^+ - \bar{H}$ subsystem and atom B as well as the interaction of B with the (quantum) radiation field.

A. Jacob, *Relativistic Effects in Interatomic Ionization Processes and Formation of Antimatter Ions in Interatomic Attachment Reactions*,
https://doi.org/10.1007/978-3-658-43891-3_8

In addition, we have considered the formation of positive ions of antihydrogen via two nonradiative three-body reactions. The first reaction is electron-assisted three-body attachment, $e^- + e^+ + \bar{H} \rightarrow e^- + \bar{H}^+$. It is driven by the interaction between the incident electron and positron resulting in the attachment of the positron whereas the electron carries away the energy release. The second nonradiative reaction is positron-assisted three-body attachment, $e^+ + e^+ + \bar{H} \rightarrow e^+ + \bar{H}^+$. Here, the driving force is the positron-positron interaction leading to the attachment of one of the incident positrons while the other takes away the released energy.

First, we have performed a comparison between the radiative mechanisms SRA and 2CDA and the nonradiative mechanism 3BAe, where the 2CDA involves Cs as atomic species B with the $6\,^2S_{1/2} \rightarrow 6\,^2P_{3/2}$ dipole transition in Cs. Our results show that at low incident energies ($\lesssim 10^{-2}$ eV) of the positron and electron with respect to the antihydrogen atom, the rate for the 3BAe can be orders of magnitude larger than the rate for SRA and 2CDA. The situation changes at incident positron energies $\varepsilon_{k_p} \simeq 1$ eV, which are most favourable for the radiative attachment mechanisms, with SRA and 2CDA now strongly dominating the 3BAe. In particular, close to the two-center resonance at $\varepsilon_{k_{p,res}} \approx 0.71$ eV, the rate for 2CDA not only greatly exceeds the rate for 3BAe but also that for SRA.

Next, we have compared the radiative mechanisms SRA and LIRA in the range of incident positron energies $\varepsilon_{k_p} \simeq 1$ eV. Here, the LIRA can, under certain conditions (including a relatively small frequency and short time duration of the laser pulse), markedly outperform the SRA.

Further, we have seen that over the whole range of incident positron energies under consideration, the 3BAp has vanishingly small rates compared to the 3BAe. This was attributed to a relatively large spatial separation of the two incident positrons in the 3BAp reaction, caused by their mutual repulsion, which greatly reduces the probability for the attachment to occur.

Besides, in order to have a more complete treatment of $\bar{H}^+$ formation mechanisms, we have compared the 3BAe with the charge exchange collision $Ps + \bar{H} \rightarrow \bar{H}^+ + e^-$, in which the bound positron in positronium is captured by antihydrogen. At the very low relative energies where the 3BAe is most efficient, the charge exchange collision is energetically not allowed if the incident Ps is in the ground state ($n = 1$) or first excited states ($n = 2$). However, considering Ps to be initially in excited states with $n = 3$, the rate for $\bar{H}^+$ formation in $\bar{H} - Ps$ collisions becomes relatively large, where the 3BAe rate will become comparable to this rate starting with electron densities of $n_e \gtrsim 10^{14}$ cm^{-3}.

The 2CDA and 3BAe both proceed in environments where antimatter is embedded in matter. Nevertheless, we have concluded that 2CDA and 3BAe are essentially not influenced by annihilation or other processes which involve the interaction between matter and antimatter.

To conclude this study on the formation of $\bar{\mathrm{H}}^+$ via radiative and nonradiative attachment of e^+ to $\bar{\mathrm{H}}$, we take a brief outlook on the experimental realization of the considered attachment mechanisms. The overwhelming majority of antihydrogen atoms produced in the antihydrogen experiments at CERN are in a broad range of (highly excited) Rydberg states and the main challenge to date is the efficient de-excitation of the formed antiatoms to the ground state. Therefore, the production of substantial amounts of $\bar{\mathrm{H}}^+$ ions via the discussed attachment mechanisms, which rely on antihydrogen being in the ground state, is currently not feasible [151]. However, there is an ongoing work on efficiently de-exciting Rydberg antihydrogen to its ground state (see, e.g. [152, 153]) indicating that, in the near future, the radiative and nonradiative attachment mechanisms considered in this work could be realized in an experiment.

Bibliography

1. J. Eichler and W. E. Meyerhof, *Relativistic Atomic Collisions* (Academic Press, Cambridge, MA, 1995).
2. J. Eichler, *Lectures on Ion-Atom Collisions* (Elsevier, Amsterdam, 2005).
3. A. B. Voitkiv, B. Najjari, R. Moshammer, and J. Ullrich, Phys. Rev. A **65**, 032707 (2002).
4. A. B. Voitkiv and B. Najjari, Phys. Rev. A **79**, 022709 (2009).
5. L. D. Landau and E. M. Lifshitz, *Relativistic Quantum Theory* (Pergamon Press Ltd., Oxford, 1971); see Sections 56; 57.
6. M. Y. Amusia, *Atomic Photoeffect* (Springer, New York, 1990).
7. J. W. Cooper, Phys. Rev. **128**, 681 (1962).
8. J. J. Yeh, I. Lindau, Atomic Data and Nuclear Data Tables **32**, 1, 1–155 (1985).
9. H. Hertz, Annalen der Physik **267**, 8, 983–1000 (1887).
10. W. Hallwachs, Annalen der Physik **269**, 2, 301–312 (1888).
11. A. Einstein, Annalen der Physik **322**, 6, 132–148 (1905).
12. T. H. Maiman, Nature **187**, 493–494 (1960).
13. N. Huang, H. Deng, B. Liu, D. Wang, Z. Zhao, The Innovation **2**, 2, 100097 (2021).
14. H. Geiger and E. Marsden, Proc. R. Soc. Lond. A **82**, 495–500 (1909).
15. E. Rutherford, The London, Edinburgh, and Dublin Philosophical Magazine and Journal of Science **21**, 125, 669–688 (1911).
16. J. Eichler, Physics Reports **193**, Nos. 4 & 5, 165–277 (1990).
17. J. Eichler, T. Stöhlker, *Relativistic Ion-Atom Collisions*. In: H. F. Beyer, V. P. Shevelko (eds), *Atomic Physics with Heavy Ions* (Springer, Berlin, Heidelberg, 1999).
18. O. Chuluunbaatar et. al., Phys. Rev. A **99**, 062711 (2019).
19. S. Grundmann et. al., Phys. Rev. Lett. **124**, 233201 (2020).
20. B. M. Smirnov, *Physics of Atoms and Ions* (Springer, New York, 2003); see pp. 193–195; 208–210.
21. L. Meitner, Zeitschrift für Physik, **11**, 35–54 (1922).
22. P. Auger, J. Phys. Radium **6**, 6, 205–208 (1925).
23. F. M. Penning, Naturwissenschaften **15**, 818 (1927).
24. T. E. Gallon and J. A. D. Matthew, Phys. Stat. Sol. **41**, 343 (1970).
25. D. G. Lord and T. E. Gallon, Surface Science **36**, 606 (1973).
26. J. A. D. Matthew and Y. Komninos, Surf. Sci. **53**, 716 (1975).
27. L. S. Cederbaum, J. Zobeley, and F. Tarantelli, Phys. Rev. Lett. **79**, 4778 (1997).

A. Jacob, *Relativistic Effects in Interatomic Ionization Processes and Formation of Antimatter Ions in Interatomic Attachment Reactions*,
https://doi.org/10.1007/978-3-658-43891-3

28. S. Marburger, O. Kugeler, U. Hergenhahn, and T. Möller, Phys. Rev. Lett. **90**, 203401 (2003).
29. T. Jahnke et. al., Phys. Rev. Lett. **93**, 163401 (2004).
30. T. Förster, Ann. Phys. **437**, 55–75 (1948).
31. E. A. Jares-Erijman and T. M. Jovin, Nat. Biotechnol. **21**, 1387 (2003).
32. J. Zobeley, R. Santra, and L. S. Cederbaum, J. Chem. Phys. **115**, 5076 (2001).
33. K. Gokhberg and L. S. Cederbaum, J. Phys. B: At. Mol. Opt. Phys. **42**, 231001 (2009).
34. A. Jacob, C. Müller, and A. B. Voitkiv, J. Phys. B: At. Mol. Opt. Phys. **52**, 225201 (2019).
35. T. Jahnke et. al., Chem. Rev. **120**, 11295–11369 (2020).
36. T. Jahnke, J. Phys. B **48**, 082001 (2015).
37. U. Hergenhahn, J. Electron Spectrosc. Relat. Phenom. **184**, 78 (2011).
38. V. Averbukh et. al., J. Electron Spectrosc. Relat. Phenom. **183**, 36 (2011).
39. A. Eckey, A. Jacob, A. B. Voitkiv, and C. Müller, Phys. Rev. A **98**, 012710 (2018).
40. C. Müller, A. B. Voitkiv, J. R. Lopez-Urrutia, and Z. Harman, Phys. Rev. Lett. **104**, 233202 (2010).
41. A. Jacob, C. Müller, and A. B. Voitkiv, Phys. Rev. A **100**, 012706 (2019).
42. F. Grüll, A. B. Voitkiv, and C. Müller, Phys. Rev. A **100**, 032702 (2019).
43. A. Jacob, C. Müller, and A. B. Voitkiv, Phys. Rev. A **103**, 042804 (2021).
44. B. Najjari, A. B. Voitkiv, and C. Müller, Phys. Rev. Lett. **105**, 153002 (2010).
45. A. B. Voitkiv and B. Najjari, Phys. Rev. A **82**, 052708 (2010).
46. F. Trinter et. al., Phys. Rev. Lett. **111**, 233004 (2013).
47. A. Mhamdi et. al., Phys. Rev. A **97**, 053407 (2018).
48. A. Hans et. al., J. Phys. Chem. Lett. **10**, 1078 (2019).
49. A. B. Voitkiv, C. Müller, S. F. Zhang and X. Ma, New J. Phys. **21**, 103010 (2019).
50. A. Jacob, C. Müller, and A. B. Voitkiv, arXiv:2208.09812 (2022).
51. A. B. Voitkiv, J. Phys. B **40**, 2885 (2007).
52. J. D. Jackson, *Classical Electrodynamics* (John Wiley & Sons, Inc., New York, 1962); see pp. 102; 393; 520.
53. R. Anholt, Phys. Rev. A **19**, 1004 (1979).
54. N. Tariq, N. Al Taisan, V. Singh, and J. D. Weinstein, Phys. Rev. Lett. **110**, 153201 (2013).
55. B. Friedrich, Physics **6**, 42 (2013).
56. Atomic spectra data base of the National Institute of Standards and Technology (NIST), available at https://www.nist.gov/pml/atomic-spectra-database.
57. D. A. Verner and D. G. Yakovlev, Astron. Astrophys. Suppl. Ser. **109**, 125 (1995).
58. A. Ben-Asher, A. Landau, L. S. Cederbaum, and N. Moiseyev, J. Phys. Chem. Lett. **11**, 6600 (2020).
59. J. A. C. Gallas, Phys. Rev. A **21**, 1829 (1980).
60. F. Grüll, A. B. Voitkiv, and C. Müller, Phys. Rev. A **102**, 012818 (2020).
61. M. Inokuti, Rev. Mod. Phys. **43**, 297 (1971).
62. J. Berkowitz, *Atomic and Molecular Photoabsorption: Absolute Total Cross Sections* (Academic Press, Cambridge, MA, 2001).
63. Y. Kim, J. P. Santos, and F. Parente, Phys. Rev. A **62**, 052710 (2000).
64. Y. Kim and M. Inokuti, Phys. Rev. **175**, 176 (1968).
65. M. Inokuti and Y. Kim, Phys. Rev. **186**, 100 (1969).

66. A. Kay et. al., Science **281**, 679 (1998).
67. G. Öhrwall et. al., Phys. Rev. Lett. **93**, 173401 (2004).
68. R. Moshammer et. al., Phys. Rev. Lett. **77**, 1242 (1996).
69. R. Moshammer et. al., Phys. Rev. Lett. **79**, 3621 (1997).
70. N. Stolterfoht et. al., Phys. Rev. Lett. **80**, 4649 (1998).
71. D. L. Burke et. al., Phys. Rev. Lett. **79**, 1626 (1997).
72. A. B. Voitkiv, B. Najjari, and J. Ullrich, Phys. Rev. Lett. **94**, 163203 (2005).
73. A. B. Voitkiv, B. Najjari, and J. Ullrich, Phys. Rev. Lett. **103**, 193201 (2009).
74. E. Fermi, Z. Phys. **29**, 315 (1924).
75. C. F. von Weizsäcker, Z. Phys. **88**, 612 (1934).
76. E. J. Williams, Kgl. Danske Videnskab. Selskab. Mat.-fys. Medd. **13**, No. 4 (1935).
77. C. A. Bertulani and G. Baur, Phys. Rep. **163**, 299 (1988).
78. G. Baur, K. Hencken and D. Trautmann, Phys. Rep. **453**, 1 (2007).
79. R. Santra, J. Zobeley, L. S. Cederbaum, and N. Moiseyev, Phys. Rev. Lett. **85**, 4490 (2000).
80. J. L. Hemmerich, R. Bennett and S. Y. Buhmann, Nat. Comm. **9**, 2934 (2018).
81. M. R. C. McDowell and J. P. Coleman, *Introduction to the theory of ion-atom collisions* (North-Holland Publishing Company, Amsterdam-London, 1970).
82. N. Stolterfoht, R. D. DuBois and R. D. Rivarola, *Electron Emission in Heavy Ion-Atom Collisions* (Springer, Berlin, Heidelberg, 1997).
83. J. H. McGuire, *Electron Correlation Dynamics in Atomic Collisions* (Cambridge University Press, 1997).
84. J. B. Delos, Rev. Mod. Phys. **53**, 287 (1981).
85. M. H. Mittleman, *Introduction to the Theory of Laser-Atom Interactions* (Springer, New York, 1993); pp. 22–24.
86. A. Voitkiv and J. Ullrich, *Relativistic Collisions of Structured Atomic Particles* (Springer, Berlin, Heidelberg, 2008); see pp. 73; 105–107.
87. M. Abramowitz and I. Stegun, *Handbook of Mathematical Functions* (Dover, New York, 1965); see Sections 9.1; 9.2; 9.6; 13.1; 15.1; 15.2.
88. F. Grüll, A. B. Voitkiv, and C. Müller, Phys. Rev. Res. **2**, 033303 (2020).
89. J. B. Mann and W. R. Johnson, Phys. Rev. A **4**, 41 (1971).
90. P. A. M. Dirac, Proc. Roy. Soc. A **133**, 60 (1931).
91. C. D. Anderson, Science **76**, 238 (1932).
92. C. D. Anderson, Phys. Rev. **43**, 491 (1933).
93. O. Chamberlain, E. Segrè, C. Wiegand, T. Ypsilantis, Phys. Rev. **100**, 947 (1955).
94. E. Segrè, C. E. Wiegand, Sci. Am. **194** (6), 37 (1956).
95. G. Baur et. al., Phys. Lett. B **368**, 251 (1996).
96. G. Blanford et. al., Phys. Rev. Lett. **80**, 3037 (1998).
97. M. Amoretti et. al., Nature **419**, 456 (2002).
98. S. Schippers et. al., J. Phys. B: At. Mol. Opt. Phys. **52**, 171002 (2019).
99. M. H. Holzscheiter, M. Charlton, and M. M. Nieto, Phys. Rep. **402**, 1 (2004).
100. C. H. Storry et. al., Phys. Rev. Lett. **93**, 263401 (2004).
101. The website of the ALPHA Collaboration is available at https://alpha.web.cern.ch/.
102. G. Andresen et. al., Nature **468**, 673 (2010).
103. The ALPHA Collaboration, Nat. Phys. **7**, 558 (2011).
104. C. Amole et. al., Nature **483**, 439 (2012).

105. The ALPHA Collaboration and A. E. Charman, Nat. Commun. **4**, 1785 (2013).
106. C. Amole et. al., Nat. Commun. **5**, 3955 (2014).
107. M. Ahmadi et. al., Nature **557**, 71 (2018).
108. T. A. Wagner, S. Schlamminger, J. H. Gundlach and E. G. Adelberger, Class. Quantum Grav. **29**, 184002 (2012).
109. The website of the AEGIS experiment is available at https://aegis.web.cern.ch/.
110. M. Doser et. al., Class. Quantum Grav. **29**, 184009 (2012).
111. The website of the GBAR experiment is available at https://gbar.web.cern.ch/.
112. P. Perez and Y. Sacquin, Class. Quantum Grav. **29**, 184008 (2012).
113. C. M. Keating, M. Charlton and J. C. Straton, J. Phys. B: At. Mol. Opt. Phys. **47**, 225202 (2014).
114. C. M. Keating, K. Y. Pak and J. C. Straton, J. Phys. B: At. Mol. Opt. Phys. **49**, 074002 (2016).
115. A. R. Swann, D. B. Cassidy, A. Deller, and G. F. Gribakin, Phys. Rev. A **93**, 052712 (2016).
116. P. Comini and P-A Hervieux, J. Phys.: Conf. Ser. **443**, 012007 (2013).
117. P. Comini and P-A Hervieux, New J. Phys. **15**, 095022 (2013).
118. P. Comini, P-A Hervieux and F. Biraben, Hyperfine Interact. **228**, 159 (2014).
119. D. A. Cooke, A. Husson, D. Lunney and P. Crivelli, EPJ Web of Conferences **181**, 01002 (2018).
120. A. Jacob, S. F. Zhang, C. Müller, X. Ma, and A. B. Voitkiv, Phys. Rev. Research **2**, 013105 (2020).
121. A. Jacob, C. Müller, and A. B. Voitkiv, Phys. Rev. A **104**, 032802 (2021).
122. A. B. Voitkiv, N. Grün, and W. Scheid, J. Phys. B **32**, 101 (1999).
123. D. Madison et. al., J. Phys. B **35**, 3297 (2002).
124. A. B. Voitkiv, Phys. Rev. A **95**, 032708 (2017).
125. P. Birk et. al., J. Phys. B **53**, 124002 (2020).
126. Y. Yamaguchi, Phys. Rev. **95**, 1628 (1954).
127. Y. Hahn, Rep. Prog. Phys. **60**, 691 (1997).
128. P. Beiersdorfer, Annu. Rev. Astron. Astrophys. **41**, 343 (2003).
129. J. Eichler and T. Stöhlker, Phys. Rep. **439**, 1 (2007).
130. A. Müller, Adv. At. Mol. Opt. Phys. **55**, 293 (2008).
131. R. K. Janev and H. van Regemorter, Astron. Astrophys. **37**, 1 (1974).
132. B. M. McLaughlin, H. R. Sadeghpour, P. C. Stancil, A. Dalgarno, and R. C. Forrey, J. Phys.: Conf. Ser. **388**, 022034 (2012).
133. I. S. Gradshteyn and I. M. Ryzhik, *Tables of Integrals, Series and Products* (Elsevier, Amsterdam, 2007); p. 1107.
134. C. M. Keating, *Using Strong Laser Fields to Produce Antihydrogen Ions* (Dissertation, Portland State University, 2018).
135. A. Jacob, Master's thesis, Heinrich-Heine-Universität Düsseldorf (2019).
136. A. Nordsieck, Phys. Rev. **93**, 785 (1954).
137. L. D. Landau and E. M. Lifshitz, *Quantum Mechanics* (Pergamon Press Ltd., Oxford, 1965); see Sections 19; 134.
138. J. Fajans and C. M. Surko, Phys. Plasmas **27**, 030601 (2020).
139. T. Mohamed, Phys. Lett. A **382**, 2459 (2018).
140. A. Wolf, Hyperfine Int. **76**, 189 (1993).

141. A. Müller and A. Wolf, Hyperfine Int. **109**, 233 (1997).
142. M. Amoretti et. al. (ATHENA Collaboration), Phys. Rev. Lett. **97**, 213401 (2006).
143. T. Yamashita, Y. Kino, E. Hiyama, S. Jonsell and P. Froelich, New J. Phys. **23**, 012001 (2021).
144. H. J. Ache, Angew. Chem. internat. Edit. **11**, 179 (1972).
145. N. Sinha, S. Singh, and B. Antony, J. Phys. B: At. Mol. Opt. Phys. **51**, 015204 (2018).
146. A. A. Kernoghan, M. T. McAlinden, and H. R. J. Walters, J. Phys. B: At. Mol. Opt. Phys. **29**, 3971 (1996).
147. J. Mitroy, M. W. J. Bromley, and G. Ryzhikh, J. Phys. B: At. Mol. Opt. Phys. **32**, 2203 (1999).
148. K. R. Lang, *Astrophysical Formulae* (Springer, Berlin, 1974); see Section 4.5.1.6.
149. S. Jonsell, Philos. Trans. R. Soc. A **376**, 2116 (2018).
150. Y. K. Ho, Chin. J. Phys. **35**, 97 (1997).
151. Private correspondence with Michael Doser who is the spokesperson of AEgIS, one of the antihydrogen experiments at CERN (see, e.g. https://aegis.web.cern.ch/home.html) (2021).
152. D. Comparat and C. Malbrunot, Phys. Rev. A **99**, 013418 (2019).
153. M. Vieille-Grosjean et. al., Eur. Phys. J. D **75**, 27 (2021).
154. J. R. Swanson and L. Armstrong, Jr., Phys. Rev. A **15**, 661 (1977).
155. D. J. Kennedy and S. T. Manson, Phys. Rev. A **5**, 227 (1972).
156. T. N. Chang and T. Olsen, Phys. Rev. A **23**, 2394 (1981).

Printed in the United States
by Baker & Taylor Publisher Services